Enacted Mathematics Curriculum

A Conceptual Framework and Research Needs

Enacted Mathematics Curriculum

A Conceptual Framework and Research Needs

edited by

Denisse R. Thompson
University of South Florida

and

Zalman Usiskin
University of Chicago

Information Age Publishing, Inc.
Charlotte, North Carolina • www.infoagepub.com

Library of Congress Cataloging-in-Publication Data

CIP data for this book can be found on the Library of Congress website http://www.loc.gov/index.html

ISBNs: Paperback: 978-1-62396-583-9
ISBNs: Hardcover: 978-1-62396-584-6
ISBNs: Ebook: 978-1-62396-585-3

Chapters in this volume are based on work done at the *Conference on Research on the Enacted Mathematics Curriculum*, funded by the National Science Foundation (DRL 0946433), and held at the University of South Florida, November 4-6, 2010. All opinions are those of the individual authors and do not necessarily represent the views of the Foundation.

Copyright © 2014 IAP–Information Age Publishing, Inc.

All rights reserved. No part of this publication may be reproduced, stored in a retrieval system, or transmitted in any form or by any electronic or mechanical means, or by photocopying, microfilming, recording or otherwise without written permission from the publisher.

Printed in the United States of America

CONTENTS

Acknowledgements . . . . *vii*

Preface . . . . *ix*

1. The Enacted Curriculum as a Focus of Research
   Gabriel Cal and Denisse R. Thompson *1*

2. Examining Variations in Enactment of a Grade 7 Mathematics Lesson by a Single Teacher: Implications for Future Research on Mathematics Curriculum Enactment
   Mary Ann Huntley and Daniel J. Heck *21*

3. Influence of Mathematics Curriculum Enactment on Student Achievement
   Patricia D. Hunsader and Denisse R. Thompson *47*

4. Teachers' Knowledge and the Enacted Mathematics Curriculum
   Ji-Won Son and Sharon L. Senk *75*

5. Instruments for Studying the Enacted Mathematics Curriculum
   Steven W. Ziebarth, Nicole L. Fonger, and James L. Kratky *97*

6. Conceptualizing the Enacted Curriculum in Mathematics Education
   Janine T. Remillard and Daniel J. Heck *121*

7. Recommendations for Generating and Implementing a Research Agenda for Studying the Enacted Mathematics Curriculum
   Kathryn B. Chval, Iris R. Weiss, and Rukiye Didem Taylan *149*

Postscript
   Zalman Usiskin *177*

Conference Agenda.. *181*
Conference Participants *185*
About the Authors.. *189*

ACKNOWLEDGEMENTS

This book would not have been possible without the support of a number of individuals and institutions. First, thanks to the National Science Foundation and initial program officer, John "Spud" Bradley, for funding the *Conference on Research on the Enacted Mathematics Curriculum* (DRL 0946433) held at the University of South Florida (Tampa) in November 2010. Without support to hold the conference, there would not have been the impetus for developing this book.

Second, thanks go to the members of the Steering Committee who provided guidance in planning the conference: Kathryn B. Chval, Daniel J. Heck, Mary Ann Huntley, Janine T. Remillard, Sharon L. Senk, Iris R. Weiss, and Steven W. Ziebarth. Each of these members was willing to write a chapter for this volume, and in some cases, also mentor early-career researchers as part of their work to develop the chapter. Thanks also to Gabriel Cal for his logistics work during the conference and his work in summarizing notes from discussions among various groups.

As the Principal Investigator of the NSF grant, I am very appreciative that Zalman Usiskin agreed to co-edit this volume with me. Professor Usiskin's long history with curriculum development and careful editing of chapters was instrumental in ensuring a coherent volume that I hope will be useful to curriculum researchers at many levels. Finally, thanks are extended to George Johnson, President of Information Age Publishing, for his interest in publishing this volume to make the work of the conference available to the wider mathematics education community.

Denisse R. Thompson

PREFACE

Throughout the 20th century, professional organizations and governmental agencies periodically formed committees to recommend and, in some cases, mandate changes in the mathematics content that is taught in schools or in the manner in which mathematics is taught. Justification for these changes arose from changes in mathematics itself, changes in beliefs about how students learn, and changes in society.

The current scene in the United States, dominated by the *Common Core State Standards for Mathematics* (Council of Chief State School Officers, 2010) and its variants, can be traced back to the 1989 *Curriculum and Evaluation Standards for School Mathematics* report of the National Council of Teachers of Mathematics (NCTM), which recommended many changes in content. This report was followed by a second NCTM report two years later, *Professional Standards for Teaching Mathematics*, which recommended changes in the ways in which mathematics was taught. Together these documents have come to be termed the *NCTM Standards*.

The importance of these documents was quickly demonstrated by the decision of the National Science Foundation (NSF) to fund multiple curriculum development projects leading to new curriculum materials at the elementary, middle, and high school levels. The NSF recognized the importance of curriculum materials to the teaching and learning of mathematics. As Begle (1973) noted with regards to comparison studies in the new math era,

> The textbook has a powerful influence on what students learn.... The evidence indicates that most student learning is directed by the text rather than the teacher. This is an important finding, since the content of the text is a variable that we can manipulate. (p. 209)

Indeed, as noted by Valverde, Bianchi, Wolfe, Schmidt, and Houang (2002), textbooks are a universal element of schools around the world: "Perhaps only students and teachers themselves are a more ubiquitous element of schooling than textbooks.... They represent school disciplines to students. They translate a country's curriculum policies into such representations" (p. 1). The NSF funding required research comparing the new curricular materials with existing materials. Most often, new and old materials are compared by how well students using them perform on tests. However, any causality that can be inferred about curricular materials on the basis of tests assumes that the teachers in the study taught in a way that is consistent with the materials they were using. In the case of the materials developed with NSF funding, this faithfulness to the materials meant that a teacher might have to change both the content that was being taught and the manner(s) in which that content was delivered.

Yet evidence exists from numerous research studies that teachers often enacted or implemented the same curriculum in different ways (Grouws & Smith, 2000; Kilpatrick, 2003; Lambdin & Preston, 1995; National Research Council [NRC], 2004; Snyder, Bolin, & Zumwalt, 1992; Spillane & Zeuli, 1999; Thompson & Senk, 2010). Kilpatrick (2003) commented on this phenomenon:

> Two classrooms in which the same curriculum is supposedly being "implemented" may look very different; the activities of teacher and students in each room may be quite dissimilar, with different learning opportunities available, different mathematical ideas under consideration, and different outcomes achieved. (p. 473)

There are many reasons for a teacher to deviate from the teaching recommended in the materials in use in his or her classroom. The teacher may feel that the book does not do an adequate job with the content, may be uncomfortable with the content (and thus skip it), may wish to adapt the content to the students in the particular class, may have a favorite way of approaching the content, or may not be comfortable with the kind of teaching recommended in the materials. Whatever the reason, the variety of enactments of the curriculum has led many researchers to study how curriculum materials are enacted within the classroom (see Gehrke, Knapp, & Sirotnik, 1992; Remillard, 2005; Remillard, Herbel-Eisenmann, & Lloyd, 2009). This book attempts to shed additional insight into the issues related to how mathematics curriculum materials are enacted as part of classroom instruction.

The importance of conceptualizing and measuring use of mathematics curriculum materials, together with influences on student achievement, received renewed attention in the last decade with calls for scientifically-based research (NRC, 2002; No Child Left Behind, 2002) and criteria for evaluating curricular effectiveness (NRC, 2004). However, just noting which textbook is used for instruction provides insufficient information about its effectiveness because of the potential for differences in the nature and quality of its enactment across teachers. As the NRC (2004) committee noted,

> [Curricular] evaluations should present evidence that provides reliable and valid indicators of the extent, quality, and type of the implementation of the materials. At a minimum, there should be documentation of the extent of coverage of curriculum material (what some investigators referred to as "opportunity to learn"). (p. 194)

Despite this interest in issues related to mathematics curriculum enactment, there has not been a systematic accumulation of knowledge that can guide policy and practice. Substantial progress in this area will require a conceptual model, a research agenda, valid and reliable tools, and a network of mathematics curriculum researchers with the capacity to conduct and synthesize the required research. To assist in this process, the National Science Foundation provided support for a *Conference on Research on the Enacted Mathematics Curriculum* (DRL 0946433) held at the University of South Florida (Tampa) in November 2010. The conference had three main goals: (1) to develop a conceptual model for research on mathematics curriculum enactment; (2) to generate a set of priority research questions for studying the enacted mathematics curriculum; and (3) to describe instruments needed to pursue the priority research questions. The conference provided an opportunity for discussions and interactions among experts in mathematics curriculum, including senior researchers and those just embarking on careers in this area, people who have been involved in the development of mathematics curriculum materials, and people whose perspective is primarily practitioner-oriented. Through the work of the conference, participants aimed to explicate theory on mathematics curriculum enactment, define key constructs and explain how they are expected to interact, and why.

This volume collates the work of the conference in one source and makes it available to a wide audience. However, the chapters extend beyond just a set of proceedings, and attempt to raise issues of interest to researchers engaged in studying varied aspects of curriculum.

Potential Audience for the Volume

We believe the volume has the potential to be useful to a range of researchers, from established veterans in curriculum research to new researchers in this area of mathematics education. The chapters can be used to generate conversation about researching the enacted mathematics curriculum, including similarities and differences in the variables that can and should be studied across various curricula. As such, it might be used by a curriculum project team as it outlines a research agenda for curriculum or program evaluation. It might be used as a text in a university graduate course on curriculum research and design.

The chapters in this volume are a natural complement to those in *Approaches to Studying the Enacted Mathematics Curriculum* (Heck, Chval, Weiss, & Ziebarth, 2012), also published by Information Age Publishing. While the present volume focuses on a range of issues related to researching the enacted mathematics curriculum, including theoretical and conceptual issues, the volume by Heck et al. (2012) provides insights into different instrumentations used by groups of researchers to study curriculum enactment. Thus, the chapters in that volume highlight ways that researchers have dealt with the complexity of measuring the enacted curriculum as teachers alter their adopted curriculum materials during use.

Structure of the Volume

The volume consists of seven chapters. Chapter 1 (Cal & Thompson) outlines different levels of curriculum often identified in the literature, setting the stage for subsequent chapters that investigate how those levels relate to aspects of enactment.

Chapters 2-5 focus on issues related to studying the enacted curriculum. Chapter 2 (Huntley & Heck) describes a situation in which one teacher implements the same lesson with two classes and considers how differences in implementation potentially provide different opportunities for student learning. Chapter 3 (Hunsader & Thompson) and Chapter 4 (Son & Senk) consider how student achievement and teacher knowledge, respectively, are influenced by and interact with the enacted curriculum. Both of these chapters were solicited to address issues not explicitly discussed at the conference but essential to understanding the broader context of the enacted curriculum. Chapter 5 (Ziebarth, Fonger, & Kratky) discusses instruments that currently exist for measuring curriculum enactment and highlights needs and issues within the realm of instrument development.

Chapters 6 and 7 discuss conceptual and research needs. Chapter 6 (Remillard & Heck) outlines a conceptual model for studying the enacted curriculum and illustrates how the variables analyzed in several research studies map to the framework outlined. This mapping of past research illustrates the viability of the conceptual framework for understanding the variables of interest in studying the enacted curriculum. Chapter 7 (Chval,

Weiss, & Taylan) then identifies steps that the mathematics education community will likely need to undertake to research the enacted mathematics curriculum and to build capacity for the work that needs to be done.

The postscript (Usiskin) encapsulates the content of this book and provides a metaphor for thinking about the place of the enacted curriculum in the overall curriculum scene. The endmatter includes the conference agenda, a list of participants with their affiliation at the time of the conference, and brief biographies of the chapter contributors.

REFERENCES

Begle, E. G. (1973). Some lessons learned by SMSG. *Mathematics Teacher, 66,* 207-214.

Council of Chief State School Officers. (2010). *Common core state standards for mathematics.* Washington, DC: Author. Retrieved from http://www.corestandards.org

Gehrke, N. J., Knapp, M. S., & Sirotnik, K. A. (1992). In search of the school curriculum. *Review of Research in Education, 18,* 51-110.

Grouws, D. A., & Smith, M. S. (2000). NAEP findings on the preparation and practices of mathematics teachers. In E. Silver & P. Kenney (Eds.), *Results from the seventh mathematics assessment of the national assessment of educational progress* (pp. 107-139). Reston, VA: National Council of Teachers of Mathematics.

Heck, D. J., Chval, K. B., Weiss, I. R., & Ziebarth, S. W. (Eds.). (2012). *Approaches to studying the enacted mathematics curriculum.* Charlotte, NC: Information Age Publishing.

Kilpatrick, J. (2003). What works? In S. L. Senk & D. R. Thompson (Eds.), *Standards-based school mathematics curricula: What are they? What do students learn?* (pp. 471-488). Mahwah, NJ: Lawrence Erlbaum.

Lambdin, D., & Preston, R. (1995). Caricatures in innovation: Teacher adaptation to an investigation-oriented middle school mathematics curriculum. *Journal of Teacher Education, 46(2),* 130-140.

National Council of Teachers of Mathematics. (1989). *Curriculum and evaluation standards for school mathematics.* Reston, VA: Author.

National Council of Teachers of Mathematics. (1991). *Professional standards for teaching mathematics.* Reston, VA: Author.

National Research Council. (2002). *Scientific research in education.* Committee on Scientific Principles for Education Research (R. J. Shavelson & L. Towne, Eds.). Washington, DC: National Academy Press.

National Research Council. (2004). *On evaluating curricular effectiveness: Judging the quality of K-12 mathematics evaluations.* Committee for a Review of the Evaluation Data on the Effectiveness of NSF-Supported and Commercially Generated Mathematics Curriculum Materials. J. Confrey & V. Stohl (Eds.), Mathematical Sciences Education Board, Center for Education, Division of Behavioral and Social Sciences and Education. Washington, DC: National Academies Press.

No Child Left Behind Act of 2001, Public Law. No. 107-110, 115 Stat. 1425 (2002).
Remillard, J. T. (2005). Examining key concepts in research on teachers' use of mathematics curricula. *Review of Educational Research*, *75*, 211-246.
Remillard, J. T., Herbel-Eisenmann, B. A., & Lloyd, G. M. (Eds.). (2009). *Mathematics teachers at work: Connecting curriculum materials and classroom instruction*. New York, NY: Routledge.
Snyder, J., Bolin, F., & Zumwalt, K. (1992). Curriculum implementation. In P. Jackson (Ed.), *Handbook of research on curriculum* (pp. 402-435). New York, NY: Macmillan.
Spillane, J. P., & Zeuli, J. S. (1999). Reform and teaching: Exploring patterns of practice in the context of national and state mathematics reforms. *Educational Evaluation and Policy Analysis*, *21*(1), 1-27.
Thompson, D. R., & Senk, S. L. (2010). Myths about curriculum implementation. In B. Reys, R. Reys, & R. Rubenstein (Eds.), *K-12 curriculum issues* (pp. 249-263). Reston, VA: National Council of Teachers of Mathematics.
Valverde, G. A., Bianchi, L. J., Wolfe, R. G., Schmidt, W. H., & Houang, R. T. (2002). *According to the book: Using TIMSS to investigate the translation of policy into practice through the world of textbooks*. Dordrecht, Netherlands: Kluwer.

CHAPTER 1

THE ENACTED CURRICULUM AS A FOCUS OF RESEARCH

Gabriel Cal and Denisse R. Thompson

Various researchers have conceptualized levels of curriculum, depending on the aims and purposes of the research about that curriculum. Some of these levels relate to written curriculum materials, such as textbooks, and their classroom implementation and use; others focus on aspects of curriculum that inform textbook development while still others focus on measures of curriculum effectiveness. In the present chapter, we highlight critical aspects of the literature identifying these levels of curriculum and attempt to situate potential research on the enacted curriculum within a broad spectrum of curriculum research. In the process, we articulate a rationale for focusing research on the enacted curriculum.

INTRODUCTION

Mathematics curriculum materials, as represented by textbooks, play a prominent role in mathematics classrooms around the world. Indeed, textbooks have been described as artifacts that "exert probabilistic influences on the educational opportunities that take place in the classrooms

Enacted Mathematics Curriculum: A Conceptual Framework and Research Needs, pp. 1–19
Copyright © 2014 by Information Age Publishing
All rights of reproduction in any form reserved.

in which they are used" (Valverde, Bianchi, Wolfe, Schmidt, & Houang, 2002, p. 2). In particular, it is typically through textbooks that national curriculum recommendations are translated from abstract statements to more practical outlines for teachers and students. Valverde et al. point out that textbooks are "mediators between the intentions of the designers of curriculum policy and the teachers that provide instruction in classrooms" (p. 2). Given the important role that textbooks play in student learning of mathematics, it is essential to understand issues surrounding research related to them.

One important issue is how teachers enact or implement the curriculum reflected in their textbook within their classroom—what is often identified as the *enacted* or *implemented* curriculum. The enacted mathematics curriculum interacts with numerous elements of the educational system, from curriculum frameworks at a state or district level to adopted textbook materials at a school or classroom level to assessments for accountability purposes at a student level. These interactions are described in various documents using a myriad of definitions and descriptions of curriculum. The purpose of this chapter is to situate the enacted mathematics curriculum within this broader literature on curriculum and on curriculum research and to set the stage for the subsequent chapters related to researching the enacted curriculum. So, in this chapter, we start by delineating conceptualizations related to different levels of curriculum, applicable not only to mathematics but across a broad spectrum of disciplines. We then discuss some of the types of research that have been conducted at these different levels. We close with a discussion of specific issues related to researching the enacted mathematics curriculum, including variables of interest and a rationale for more extensive research of the enacted curriculum as represented by the subsequent chapters in this volume.

CONCEPTUALIZATIONS OF CURRICULUM

Schmidt and colleagues characterize curriculum as "the most fundamental structure" for the educational experiences of both teachers and students, describing it as a "'skeleton' that gives shape and direction to mathematics instruction in educational systems around the world" (Schmidt, McKnight, Valverde, Houang, & Wiley, 1997, p. 4). These daily experiences in the classroom are shaped based on visions and goals by policy leaders of what education they want students to have, by ideas of curriculum developers and teachers of how to create learning experiences for students, and by organized opportunities provided within the structure of schools to lead to student learning. The details of the classroom environment often mask the fact that it is the curriculum with its planned instructional activities that are the means to implement the vision and expectations of educational authorities (Schmidt et al., 1997).

Non-Mathematics Perspectives on Curriculum Levels

Since at least the 1970s, educational researchers have provided different perspectives on curriculum as it moves from a set of intentions to a set of interactions. One of the early descriptions applicable across a range of disciplines came out of the work by John Goodlad and his colleagues in *A Study of Schooling* (Goodlad, Sirotnik, & Overman, 1979). From that work, Klein, Tye, and Wright (1979) cite five perspectives or dimensions of curriculum: the *ideal*, the *formal*, the *instructional*, the *operational*, and the *experiential*.

The *ideal curriculum* represents "the beliefs, opinions, and values of the scholars in the disciplines and in schooling regarding what ought to be included in the curriculum and how it ought to be developed" (Klein, Tye, & Wright, 1979, p. 244). The *formal curriculum* is a set of written statements, such as state or district guidelines, department syllabi, or textbook resources, that detail what should be taught. The formal curriculum is translated into the *instructional curriculum* as teachers bring their beliefs, attitudes, and values about the curriculum into play in the classroom with the *operational curriculum* referring to what actually happens throughout the teaching-learning process. The instructional and operational curriculum interact to create the *experiential curriculum* which considers both students' perceptions of what curriculum is offered as well as what they actually learn.

Curriculum Levels in International Studies

When the Second International Mathematics Study [SIMS] began its planning work in the late 1970s, discussions centered on three areas: curriculum, such as the content taught and with what emphasis; the classroom, including the actions of teachers during instruction; and the end results of classroom instruction, specifically what students learn and their attitudes toward mathematics. This led to a conceptual model considering three levels of curriculum: the *intended curriculum*; the *implemented curriculum*; and the *attained curriculum*. In SIMS, the intended curriculum provided a measure of whether the mathematical content of the tested items was represented in official curriculum documents and in approved textbooks, the implemented curriculum provided a perspective from teachers as to whether they had taught the content assessed on the items, and the attained curriculum reflected students' level of achievement on the items at the beginning and end of the academic year (Travers, 1992).

This tri-partite model of *intended*, *implemented*, and *attained* curriculum continued as the initial conceptual research design of the Trends in Mathematics and Science Study [TIMSS],[1] but with some broadening of these terms to go beyond just the content of the items (Schmidt et al., 1997; Valverde et al., 2002). Within the TIMSS model, the *intended* curriculum occurs at the educational system level and is represented in national policies and official documents that indicate the values and educational objectives desired of all students at various grade levels. The teachers' intentions and objectives and how these play out in classroom activities comprise the *implemented* curriculum. As in *A Study of Schooling*, the *attained* curriculum for TIMSS represents what students have learned, including knowledge of content as well as the ability to engage in various performance expectations, such as using routine procedures, investigating and problem solving, mathematical reasoning, and communication.

The designers of TIMSS recognized that textbooks serve to link the intended curriculum and the actual implemented curriculum, and in doing so are mediators between these two levels. Hence, textbooks should represent a fourth level—the *potentially implemented* curriculum—linking aims and intentions with classroom reality. Given this perspective, researchers added this level of curriculum to the original tri-partite model (Robitaille, Schmidt, Raizen, McKnight, Britton, & Nicol, 1993; Schmidt et al., 1997; Valverde et al., 2002).

Curriculum Perspectives from Other Researchers

The curriculum models identified by Klein et al. (1979) and the researchers associated with SIMS and TIMSS have been adapted and modified by other curriculum researchers, with additional levels added to acknowledge various stages in the process from national curriculum recommendations to students' learning. For instance, Burkhardt, Fraser, and Ridgway (1990) and Clements (2002) characterized six levels of curriculum: the *ideal*, *available*, *adopted*, *implemented*, *achieved*, and *tested*. These classifications proceed from what curriculum developers intend (the ideal curriculum) to what teachers may use (available and adopted) to what teachers actually do use (implemented) to what is assessed (tested) and what students learn (achieved).

Similarly, MacNab (2000) classified curriculum as *intended*, *implemented*, and *experienced*. In this scheme, the intended curriculum represents what the curriculum designers planned to accomplish. The implemented curriculum describes how the teachers use the curriculum as part of their instruction. The experienced curriculum corresponds to how students undergo, understand, and interpret the curriculum, which can be influenced by their prior knowledge and experiences.

These characterizations of curriculum relate to stages through which a curriculum progresses from the intentions of designers or policymakers to implementation in terms of a written textbook or classroom instruction to assessments at the classroom or district/state level. Some educators (e.g., Glatthorn, 1999; Porter, 2004) use *intended* curriculum to refer to those documents, such as curriculum frameworks, that are developed by state educational agencies and define what students need to know at particular grade levels. Then, the *written* curriculum, sometimes referenced as the *textbook* curriculum (Tarr, Chávez, Reys, & Reys, 2006), is defined as "not only the content of courses but also the sequence of topics and quite often the pedagogical strategies to employ in teaching them" (Venezky, 1992, p. 439). This written curriculum provides a day-to-day plan that teachers can use to implement the curriculum in their classrooms. The *assessed* curriculum, sometimes called the *tested* curriculum (Glatthorn, 1999), then focuses on the content of exams, including classroom unit tests which teachers use to evaluate students and state-level tests used to hold schools accountable for student learning (Porter, 2004).[2]

Hjalmarson (2008) argues that issues regarding curriculum are complex and involve more than just a resource used by teachers and students. A curriculum carries interpretations, expectations, and cultural values that are addressed in a systematic manner that includes pedagogical interactions between teachers and students as they use materials. This view is consistent with that expressed by Clements (2002) in which curriculum is a means to guide students' acquisition of culturally-valued concepts, procedures, and dispositions.

Specific Perspectives on the Enacted Curriculum

From a curriculum enactment orientation, curriculum is viewed as the educational experiences jointly created by students and teachers. Teachers assume an active role as curriculum makers through their assessment of students' needs and subsequent improvisation and development of their own pedagogical techniques to meet these needs. In this view, the process of enacted curriculum is one of continual growth for both teachers and students (Gujarati, 2011).

According to Bouck (2008), some scholars view the enacted curriculum as more encompassing than just the operationalization of the intended curriculum, and should include teachers' decisions in implementing the written curriculum. That is, the enacted curriculum should encompass formal and informal lessons and activities as well as teachers' behaviors, student groupings, management strategies, beliefs, and classroom comments. For example, Snyder, Bolin, and Zumwalt (1992) expanded the definition of enacted curriculum to include the co-construction of educational experiences by teachers and students. They suggested that the enacted curriculum is a transactional process whereby teachers and students interact, construct, and make meaning of the curriculum and educational experiences within a given context. However, Bouck contends that this expanded definition may still be too narrow. She suggests that the enacted curriculum is a transactional process, co-constructed by the current situation and histories of teachers, students, the schools, communities, legislation and policy, and curriculum materials.

Summary

The literature cited in this section clearly indicates variability in conceptualizations of curriculum, from the number of levels described to the language used in classifying the levels. However, there is clearly evidence of overlap among the levels and in how the levels are grouped among researchers, suggesting a need to establish a common language among those researching different aspects of curriculum. The conceptual framework discussed by Remillard and Heck (Chapter 6 of this volume) is one effort toward beginning to establish such a common language.

RESEARCHING DIFFERENT LEVELS OF THE MATHEMATICS CURRICULUM

Research on mathematics curriculum, particularly as expressed in textbooks, can take a number of forms. In a survey of relevant research, Fan, Zhu, and Miao (2013) found 111 articles or other publications related to research on mathematics textbooks, with only 6 of those publications occurring before 1980 and 83 occurring since 1990, perhaps reflecting the emphasis in the U.S. on textbook research in conjunction with the mathematics reforms introduced by the *Curriculum and Evaluation Standards for School Mathematics* (National Council of Teachers of Mathematics [NCTM], 1989). The researchers classified these 111 publications into four categories: the role of textbooks, including philosophical articles related to design; textbook analysis and comparison, including features as well as similarities and differences among textbooks; textbook use, including how textbooks are used in classrooms to influence instruction and learning; and other areas, including the relationship between textbooks and achievement. They found that 34% of the publications focused on textbook analysis, 29% on textbook comparison, and 25% on textbook use. The other 12% focused on other areas, including the role of the textbook.

The categories identified by Fan et al. as having the most focus by curriculum researchers correspond in many ways to the potentially implemented and enacted (i.e., implemented) curriculum described in the previous section. Perhaps this is to be expected, because these are the levels most directly applicable in the classroom. Nevertheless, some research has been conducted at each of the main levels of curriculum previously identified. Here, we provide some perspective, although not exhaustive, of research at these levels, identifying some areas of interest at each level.

Research on the Intended Curriculum

In the Second International Mathematics Study, curriculum analysis of the intended curriculum was undertaken to understand the curricular contexts in which teaching and learning take place. Curriculum analysis focused on the educational system as a whole (e.g., school organization, selectivity, educational goals). Sources of data included a questionnaire, national case studies, and a detailed investigation of five mathematical topics (introductory algebra, geometry, measurement, fractions, and ratios/proportions/percent) to determine the articulation of those topics between the curriculum and the classroom (Travers, 1992).

Likewise, researchers involved with the Trends in Mathematics and Science Study analyzed national curriculum documents from numerous countries to determine grade levels at which particular mathematics topics are expected to be taught (Schmidt et al., 1997; Schmidt, Houang, & Cogan, 2002). They found differences in the grade level placement of topics, the duration within the grades at which a topic was taught, and the specificity at which content and performance expectations were made. These differences highlight variation in the educational visions across countries about what students are expected to learn and when.

Using similar methodology, Reys and colleagues (2006) analyzed the curriculum documents for grades K-8 of those U.S. states with such documents, indicating the grades at which topics were introduced and the grade at which the topic was expected to be mastered. Their work highlighted considerable variability in the expected placement of topics within the curriculum across the various states. Among the recommendations arising from their work were a need to identify major foci for each grade so that students could focus on fewer learning objectives to be able to study them at a deeper level.

Research on the Textbook or the Potentially Implemented Curriculum

International, national, or state curriculum framework documents are generally the basis for the development of textbooks. In the United States, textbooks (i.e., the potentially implemented curriculum) have typically been developed by a team of authors or curriculum developers who design textbooks to reflect their interpretations of the intentions embedded in official curriculum framework documents. Teams may interpret the intended curriculum quite differently, with different design philosophies and student expectations (Hirsch, 2007). These different design perspectives may result in dissimilar opportunities for student learning.

In the design of TIMSS, researchers engaged in a detailed analysis of textbooks for the U.S. as well as many other countries, recognizing that for many U.S. teachers the textbook is essentially the curriculum (Schmidt, McKnight, & Raizen, 1997). They indicated that "without restricting what teachers *may* choose to do, they [textbooks] drastically affect what U.S. teachers are *likely* to do under the pressure of daily instruction" (p. 52, italics in the original). Researchers analyzed content, including the number of topics contained within the textbooks and the five topics emphasized the most. In addition to content analysis, researchers also investigated the extent to which the performance expectations within the TIMSS framework (knowing, using routine procedures, investigating and problem solving, mathematical reasoning, and communicating) were present, providing insight into which expectations were most emphasized in textbooks across countries.

More recently, a number of researchers have conducted detailed content analyses of U.S. textbooks. Some have focused on historical analyses of textbooks, including both content and cognitive demand of tasks (Baker, Knipe, Collins, Leon, Cummings, Blair, & Gamson, 2010; Dogbey & Kersaint, 2012; Jones & Tarr, 2007). Others have focused on comparing the content and cognitive demand of specific topics for a range of contemporary textbooks (e.g., Johnson on proportional reasoning, 2010; Pickle on statistics, 2012; Zorin on geometric transformations, 2011). Still others have focused on the extent to which mathematical processes, such as reasoning and proof, are integrated within textbooks (e.g., Stylianides' editorship of a special issue of the *International Journal of Educational Research*, in press; Thompson, Senk, & Johnson, 2012).

Researchers interested in investigating the relationship between curriculum and student achievement often argue that a first step in such research requires understanding the potential opportunities to learn available in textbooks. As noted by Begle (1973), topics have the potential to be implemented in the classroom when they exist in the textbook, and are less likely to be implemented when they are not present. By first understanding what exists within the textbook, researchers can then investigate how the textbook is implemented in the classroom and what achievement occurs as a result of that implementation.

Research on the Implemented Curriculum

The link between curriculum as represented in instructional materials (i.e., the textbook) and its influence on student learning cannot be understood without examining the curriculum as designed by teachers and enacted as part of classroom instruction. Hiebert and Grouws (2007) indicate that students' opportunities to learn are influenced by

> The emphasis teachers place on different learning goals and different topics, the expectations for learning that they set, the time they allocate for particular topics, the kinds of tasks they pose, the kinds of questions they ask and responses they accept, [and] the nature of the discussions they lead. (p. 379)

Understanding and documenting these differences in classroom instruction is the focus of research on curriculum enactment.

One aspect of the tri-partite curriculum model used in the Second International Mathematics Study focused on a study of classroom processes to provide information about the activities that occurred in classrooms as mathematics was taught (Travers, 1992). Information on classroom processes was sought through questionnaires from both teachers and students. From teachers, information was obtained about the extent to which the curriculum as posed by a national ministry, school system, or other authority outside the classroom was actually taught by the teacher; from students, information was sought about the educational level of parents, time spent on homework, and attitudes towards mathematics. In addition, information on instructional practice was collected in terms of teacher behaviors toward the subject matter, including rationales for teaching specific content, contexts of teaching, differentiation of instruction, affective issues, and reasons for student progress (Cooney, 1992). The various instructional decisions that teachers make, together with the availability of content in the written textbook and students' behaviors and attitudes, help influence students' opportunity to learn mathematics, one of the most important factors in understanding student achievement (Husen, 1967).

The interest in curriculum in the U.S. that resulted from the release of the *Curriculum and Evaluation Standards for School Mathematics* (NCTM, 1989) led to increased interest in mathematics classroom instruction and to the extent to which mathematics instruction aligned with the vision espoused in the *Standards*. As part of TIMSS, additional information was solicited about instructional practices through a study of teachers' lesson structures and beliefs, students' perceptions of their teachers' practices, and video studies of lesson instruction in several countries (Hiebert et al., 2003; Schmidt, McKnight, Cogan, Jakwerth, & Houang, 1999). These researchers suggested that U.S. teachers had more activities of short duration with less coherent and effective instruction than was true of teachers in other countries.

As new curriculum materials were developed in response to the recommendations of the NCTM *Standards*, numerous researchers began to investigate the nature of classroom instruction, believing that teachers and instruction can be and might have changed as a result of using new materials. For instance, curricula introduced in the 1990s were frequently used to introduce teachers to reform recommendations (Kon, 1994; Sykes, 1990). These curricula were generally different from what teachers had previously used, often introducing both new mathematical topics and new pedagogical approaches (Collopy, 2003; Usiskin, 2013; Wilson & Lloyd, 2000). Consequently, researchers were interested in how these materials were implemented in the classroom in relation to the philosophical stance of the curriculum designers, highlighting classroom practices consistent with or in opposition to the stance of the textbooks' authors (Remillard & Bryans, 2004; Remillard, Herbel-Eisenmann, & Lloyd, 2009; Romberg & Shafer, 2008; Schoen, Ziebarth, Hirsch, & BrckaLorenz, 2010; Tarr, Reys, Reys, Chávez, Shih, & Osterlind, 2008; Thompson, & Senk, 2010). Understanding the nature of instruction was important to understand achievement differences that might or might not be evident when students studied from different curriculum materials.

Research on the Assessed and Attained Curriculum

One issue often facing curriculum researchers is what content is appropriate to assess on tests, particularly on tests measuring student achievement across educational systems or among students using different curriculum materials with different objectives and goals (Usiskin, 2013). There can often be considerable variability in terms of items whose content teachers report having taught, within or across systems (Schmidt, Wolfe, & Kifer, 1992) or even across teachers within the same school (Thompson & Senk, 2010). Even within a single educational system (either national or state), the alignment of national goals, curriculum, and the content of national assessments may not be as strong as desired (Cal, 2011; Porter, Polikoff, Zeidner, & Smithson, 2008; Porter & Smithson, 2001; Webb,

2007). Furthermore, exploring the extent to which mathematical processes (e.g., reasoning, communication, connections, or representations) are integrated into assessments accompanying published curricula suggests that assessments used by many classroom teachers provide substantial differences in opportunities for students to demonstrate their conceptual understanding of mathematics (Hunsader, Thompson, & Zorin, 2013).

Given the variability in the assessed curriculum, it is natural to expect variability in the achieved curriculum. In the Second International Mathematics Study, there was an attempt to consider achievement in terms of growth from a pretest to a posttest and in relation to teaching practices;

> the practice of sorting students into classrooms or schools according to (apparently) some measure of prior achievement ... leads to a differentiation of curriculum that precludes the opportunities for a substantial number of students to be exposed to mathematical materials and, of course, to learn about them. (Schmidt, Wolfe, & Kifer, 1992, p. 98)

Detailed study of achievement by content subtests and curriculum emphasis continued in the detailed studies of TIMSS (Schmidt et al., 1999), with achievement better in some areas and worse in others, reflecting the fact that the curriculum does matter.

A number of studies throughout the last two decades have attempted to link student achievement to the use of particular curriculum materials or sequences when controlling for prior achievement or teaching practices (Grouws, Tarr, Chávez, Sears, Soria, & Taylan, 2013; Romberg & Shafer, 2008; Senk & Thompson, 2003; Tarr, Grouws, Chávez, & Soria, 2013; Thompson, Senk, & Yu, 2012). These studies are beginning to build models that link the varied levels of curriculum in a meaningful fashion—potentially implemented, implemented, and attained—helping to provide empirical evidence for areas of the conceptual framework outlined in Chapter 6 of this volume (Remillard & Heck) and providing a foundation to continue research along the lines of the agenda outlined in Chapter 7 (Chval, Weiss, & Taylan).

Summary

As evident throughout this section, each level of curriculum can be the focus of research efforts. Although national policy recommendations for curriculum may be difficult for individual textbook authors or teachers to influence, such individuals can exert considerable influence over the content and performance expectations of a textbook or on their own classroom instruction. Opportunities to learn content and the pedagogical nature of that learning strongly influence the opportunity to learn mathematics.

STUDYING THE ENACTED MATHEMATICS CURRICULUM: POTENTIAL VARIABLES OF INTEREST

Numerous factors mediate implementation of curriculum: a teacher's beliefs and knowledge about content, teaching, and conceptions of learning; a teacher's orientation towards curriculum, including the selection of content and instructional approach; a teacher's professional identity and community; classroom structures and norms, including the dynamics of the classroom and the nature of interactions between a teacher and his/her students; length of time for class instruction; needs of individual learners; and a school's organization and policy context (Burstein, 1992; Hiebert & Grouws, 2007; Kifer & Burstein, 1992; Stein, Remillard, & Smith, 2007). Studies related to curriculum use point to the fact that curricula are planned and implemented within the context of these and other significant factors (Remillard, 2005). In any one study, researchers can typically focus on only a small number of variables to explore areas of agreement and disagreement between curricular intentions and enactment. To understand the interplay between curriculum materials as expressed in textbooks and student achievement, much more knowledge is needed about how those materials are used as part of instruction. That is, much more research is needed on the enacted mathematics curriculum.

Given the diversity of curriculum materials developed in response to the *Curriculum and Evaluation Standards for School Mathematics* (NCTM, 1989) that ushered in the "reform" era in the U.S. and the subsequent controversies related to those materials, a committee of the National Research Council (NRC) undertook a study to investigate the effectiveness of these curricular materials. The committee recognized that curriculum implementation is complex, with many factors influencing their implementation; consequently, studies of curriculum and its effectiveness need to recognize this complexity. Nevertheless, the committee noted that studies attempting to provide judgments about a curriculum's effectiveness must include "evidence that provides reliable and valid indicators of the extent, quality, and type of the implementation of the materials. At a minimum, there should be documentation of the extent of coverage of curricula material" (NRC, 2004, p. 194). That is, the NRC report indicated that studies of curriculum effectiveness cannot focus solely on student achievement. Rather, such studies must pay attention to the enactment of the curriculum and how various factors influence that enactment.

The NRC report classified potential variables that influence implementation into three categories: *resource variables* (e.g., what resources exist to facilitate implementation); *process variables* (e.g., how implementation activities are conducted and what decisions relative to curriculum and instruction are made), and *community or cultural variables* (e.g., beliefs or expectations about teaching and learning held implicitly or explicitly by those involved in adoption or implementation of materials). In addition, the NRC report recognized that other factors mediate the quality of the implementation, such as differences in assignment of students to curriculum, the use of formative or other assessments, allocations of time and resources, influence of parent groups on instruction and curriculum, and differences in nature of instruction (NRC, 2004, pp. 44-48). Based on this work, Table 1.1 lists a number of variables that might be studied as part of research on the enacted mathematics curriculum.

Table 1.1. Potential Variables of Interest in Studying the Enacted Mathematics Curriculum

Resource Variables	Process Variables	Community/Cultural Variables
• Teacher factors: supply, credentials, and stability	• Teachers' professional community	• Beliefs of teachers about the teaching-learning-assessment cycle
• Opportunities for professional development	• Teachers' involvement in curricular decisions	• Educational and career aspirations of individuals in the community
• Length of time for instruction	• Requirements of a course	• Expectations for homework
• Class size	• Placement of students in courses	• Language proficiency
• Number of course preparations per teacher	• Nature of decision making within the school governance structure	• Student mobility
• Access to appropriate manipulatives or technology	• Frequency of assessments and nature of how the data are used	• Socioeconomic status of schooling participants (teachers and students)
• Support services to meet the needs of students		• Ethnic or racial composition of students and teachers in relation to the community
• Nature of parental involvement		• Community responses to school accountability measures
		• Parental expectations

Adapted from National Research Council (2004, p. 45)

No one study of the enacted or implemented curriculum can reasonably expect to address more than a few of these variables. As the chapters in this book illustrate, there is room for much research in this area. Chapters 2-5 each tackle an important issue in investigating some of the variables identified in Table 1.1: considering variations in classroom instruction and the implications for differences in opportunities to learn (Chapter 2); how student achievement can be influenced by differences in enactment (Chapter 3); how teacher's content and pedagogical knowledge influence their curricular enactment (Chapter 4); and what instruments exist and are still needed to study various aspects of enactment (Chapter 5). Chapters 6 and 7 then offer suggestions for future study of the enacted curriculum. Chapter 6 offers a conceptual model that might serve as one means of bringing coherency to the study of curriculum enactment; Chapter 7 provides recommendations for generating and implementing a research agenda related to this area. All of these issues related to enactment typically play out in the curriculum development and revision process.

CONCLUSION

Developing a better understanding of the research on the enacted mathematics curriculum is one goal of this volume. As noted by Usiskin (2010), "classes differ far more than would be expected by chance, and teachers and schools differ on so many different variables that a textbook or curriculum that works better in one place cannot be predicted with certainty to work better in another" (p. 36). Making predictions about curriculum effectiveness will require more causal and correlational studies (Fan, 2013) if we want to gain insight into the question, "How does curriculum A work compared to curriculum B with a given set of students and teachers in a given context?"

We believe that investigating student achievement data related to curricular materials without some insight into how those materials are enacted and what factors influence the nature of the enactment is telling only a part of the curriculum story. The complete story requires insights into what happens between the intended curriculum of state standards, the written or potentially implemented curriculum of the textbook, and the assessed curriculum. The missing link is research on the enacted curriculum. This volume is one more step in making the link explicit.

NOTES

1. TIMSS was originally called the Third International Mathematics and Science Study.
2. The reader is encouraged to read a second analysis of the various levels of curriculum given by Remillard and Heck in Chapter 6 of this volume.

REFERENCES

Baker, D., Knipe, H., Collins, J., Leon, J., Cummings, E., Blair, C., & Gamson, D. (2010). One hundred years of elementary school mathematics in the United States: A content analysis and cognitive assessment of textbooks from 1900 to 2000. *Journal for Research in Mathematics Education, 41*(4), 383-423.

Begle, E. G. (1973). Some lessons learned by SMSG. *Mathematics Teacher, 66,* 207-214.

Bouck, E. C. (2008). Exploring the enactment of functional curriculum in self-contained cross-pedagogical programs: A case study. *The Qualitative Report, 13,* 495-530.

Burkhardt, H., Fraser, R., & Ridgway, J. (1990). The dynamics of curriculum change. In I. Wirszup & R. Streit (Eds.), *Developments in school mathematics around the world* (Vol. 2, pp. 3-30). Reston, VA: National Council of Teachers of Mathematics.

Burstein, L. (1992). Studying learning, growth, and instruction cross-nationally: Lessons learned about why and why not engage in cross-national studies. In L. Burstein (Ed.), *The IEA study of mathematics III: Student growth and classroom processes* (pp. xxvii-lii). Oxford: Pergamon Press.

Cal, G. (2011). *Opportunity to learn (OTL) and the alignment of upper division mathematics learning outcomes, textbooks, and the national assessment in Belize* (Unpublished PhD dissertation). University of South Florida.

Chval, K. B., Weiss, I. R., & Taylan, R. D. (2014). Recommendations for generating and implementing a research agenda for studying the enacted mathematics curriculum. In D. R. Thompson & Z. Usiskin (Eds.), *Enacted mathematics curriculum: A conceptual framework and research needs* (pp. 149-176). Charlotte, NC: Information Age Publishing.

Clements, D. H. (2002). Linking research and curriculm development. In L. D. English (Ed.), *Handbook of internatonal research in mathematics education* (pp. 599-636). Mahwah, NJ: Lawrence Erlbaum.

Collopy, R. (2003). Curriculum materials as a professional development tool: How mathematics textbooks affected two teachers' learning. *Elementary School Journal, 103*(3), 287-311.

Cooney, T. J. (1992). Classroom processes: Conceptual considerations and design of the study. In L. Burstein (Ed.), *The IEA study of mathematics III: Student growth and classroom processes* (pp. 15-27). Oxford, England: Pergamon Press.

Dogbey, J., & Kersaint, G. (2012, February). Treatment of variables in popular middle-grades mathematics textbooks in the USA: Trends from 1957 through 2009. *International Journal of Mathematics Teaching and Learning.* Retrieved from http://www.cimt.plymouth.ac.uk/journal/

Fan, L. (2013). Textbook research as scientific research: towards a common ground on issues and methods of research on mathematics textbooks. *ZDM: The International Journal of Mathematics Education, 45,* 765-777.

Fan, L., Zhu, Y., & Miao, Z. (2013). Textbook research in mathematics education: development status and directions. *ZDM: The International Journal of Mathematics Education, 45*(5), 633-646.

Glatthorn, A. A. (1999). Curriculum alignment revisited. *Journal of Curriculum and Supervision, 15*(1), 26-34.

Goodlad, J. I., Sirotnik, K. A., & Overman, B. C. (1979). An overview of 'A Study of Schooling.' *The Phi Delta Kappan, 61*(3), 174-178.

Grouws, D. A., Tarr, J. E., Chávez, Ó., Sears, R., Soria, V., & Taylan, R. D. (2013). Curriculum and implementation effects on high school students' mathematics learning from curricula representing subject-specific and integrated content organizations. *Journal for Research in Mathematics Education, 44*(2), 416-463.

Gujarati, J. (2011). From curriculum guides to classroom enactment: Examining early career elementary teachers' orientations toward standards-based mathematics curriculum implementation. *Journal of Mathematics Education at Teachers College, 2*, 40-46.

Hiebert, J., Gallimore, R., Garnier, H., Givven, K. B., Hollingsworth, H., Jacobs, J., et al. (2003). *Teaching mathematics in seven countries: Results from the TIMSS 1999 video study.* Washington, DC: National Center for Education Statistics.

Hiebert, J., & Grouws, D. A. (2007). The effects of classroom mathematics teaching on students' learning. In F. K. Lester (Ed.), *Second handbook of research on mathematics teaching and learning* (pp. 371-404). Charlotte, NC: Information Age Publishing.

Hirsch, C. R. (Ed.). (2007). *Perspectives on the design and development of school mathematics curricula.* Reston, VA: National Council of Teachers of Mathematics.

Hjalmarson, M. A. (2008). Mathematics curriculum systems: Models for analysis of curricular innovation and development. *Peabody Journal of Education, 83*(4), 592-610.

Hunsader, P. D., Thompson, D. R., & Zorin, B. (2013, May). Engaging elementary students with mathematical processes during assessment: What opportunities exist in tests accompanying published curricula? *International Journal of Mathematics Teaching and Learning.* Retrieved from http://www.cimt.plymouth.ac.uk/journal/

Husen, T. (Ed.). (1967). *International study of achievement in mathematics: A comparison of twelve systems. Volumes I and II.* Stockholm, Sweden: Almqvist & Wiksell.

Johnson, G. J. (2010). *Proportionality in middle school mathematics textbooks* (Unpublished PhD dissertation). University of South Florida.

Jones, D. L., & Tarr, J. E. (2007). An examination of the levels of cognitive demand required by probability tasks in middle grades mathematics textbooks. *Statistics Education Research Journal, 6*(2), 4-27.

Kifer, E., & Burstein, L. (1992). Concluding thoughts: What we know, what it means. In L. Burstein (Ed.), *The IEA study of mathematics III: Student growth and classroom processes* (pp. 329-341). Oxford, England: Pergamon Press.

Klein, F. M., Tye, A. K., & Wright, F. J. (1979). A study of schooling: Curriculum. *Phi Delta Kappan, 61*(4), 244-248.

Kon, J. H. (1994). *The thud at the classroom door: Teachers' curriculum decision-making in response to a new textbook* (Unpublished doctoral dissertation). Stanford University.

MacNab, D. (2000). Raising standards in mathematics education: Values, vision, and TIMSS. *Educational Studies in Mathematics, 42*(1), 61-80.

National Council of Teachers of Mathematics. (1989). *Curriculum and evaluation standards for school mathematics*. Reston, VA: Author.

National Research Council. (2004). *On evaluating curricular effectiveness: Judging the quality of K-12 mathematics evaluations*. Committee for a Review of the Evaluation Data on the Effectiveness of NSF-Supported and Commercially Generated Mathematics Curriculum Materials. J. Confrey and V. Stohl (Eds.). Mathematical Sciences Education Board, Center for Education, Division of Behavioral and Social Sciences and Education. Washington, DC: National Academies Press.

Pickle, M. C. C. (2012). *Statistical content in middle grades mathematics textbooks* (Unpublished PhD dissertation). University of South Florida.

Porter, A. C. (2004). Curriculum assessment. Pre-publication draft to appear in J. Green, G. Camilli, & P. Elmore (Eds.), *Complementary methods for research in education*. Washington, DC: American Educational Research Association. Retrieved from http://datacenter.spps.org/uploads/curricassess.pdf.

Porter, A. C., Polikoff, M. S., Zeidner, T., & Smithson, J. (2008, Winter). The quality of content analyses of state achievement tests and content standards. *Educational Measurement: Issues and Practice*, 2-14.

Porter, A. C., & Smithson, J. L. (2001). Are content standards being implemented in the classroom? A methodology and some tentative answers. In S. H. Fuhrman (Ed.), *From the Capitol to the classroom. Standards-based reform in the States* (pp. 60-80). Chicago, IL: National Society for the Study of Education, University of Chicago Press.

Remillard, J. T. (2005). Examining key concepts in research on teachers' use of mathematics curricula. *Review of Educational Research, 75*(2), 211-246.

Remillard, J. T., & Bryans, M. B. (2004). Teachers' orientations toward mathematics curriculum materials: Implications for teacher learning. *Journal for Research in Mathematics Education, 35*(5), 352-388.

Remillard, J. T., & Heck, D. J. (2014). Conceptualizing the enacted curriculum in mathematics education. In D. R. Thompson & Z. Usiskin (Eds.), *Enacted mathematics curriculum: A conceptual framework and research needs* (pp. 121-148). Charlotte, NC: Information Age Publishing.

Remillard, J. T., Herbel-Eisenmann, B. A., & Lloyd, G. M. (Eds.). (2009). *Mathematics teachers at work: Connecting curriculum materials and classroom instruction*. New York, NY: Routledge.

Reys, B. J. (Ed.). (2006). *The intended mathematics curriculum as represented in state-level curriculum standards: Consensus or confusion?* Charlotte, NC: Information Age Publishing.

Robitaille, D. F., Schmidt, W. H., Raizen, S. A., McKnight, C. C., Britton, E., & Nicol, C. (1993). *Curriculum frameworks for mathematics and science* (TIMSS Monograph No. 1). Vancouver, Canada: Pacific Educational Press.

Romberg, T. A., & Shafer, M. C. (2008). *The impact of reform instruction on student mathematics achievement: An example of a summative evaluation of a standards-based curriculum*. New York, NY: Routledge.

Schmidt, W., Houang, R., & Cogan, L. (2002, Summer). A coherent curriculum: The case of mathematics. *American Educator*, 10-27, 47-48.

Schmidt, W. H., McKnight, C. C., Cogan, L. S., Jakwerth, P. M., & Houang, R. T. (1999). *Facing the consequences: Using TIMSS for a closer look at U.S. mathematics and science education.* Dordrecht, Netherlands: Kluwer.

Schmidt, W. H., McKnight, C. C., & Raisen, S. A. (1997). *A splintered vision: An investigation of U.S. science and mathematics education.* Dordrecht, Netherlands: Kluwer.

Schmidt, W. H., McKnight, C. C., Valverde, G. A., Houang, R. T., & Wiley, D. E. (1997). *Many visions, many aims* (Vol 1.) Dordrecht, Netherlands: Kluwer.

Schmidt, W. H., Wolfe, R. G., & Kifer, E. (1992). The identification and description of student growth in mathematics achievement. In L. Burstein (Ed.), *The IEA study of mathematics III: Student growth and classroom processes* (pp. 59-99). Oxford, England: Pergamon Press.

Schoen, H. L., Ziebarth, S. W., Hirsch, C. R. & BrckaLorenz, A. (2010). *A five-year study of the first edition of the Core-Plus Mathematics curriculum.* Charlotte, NC: Information Age Publishing.

Senk, S. L., & Thompson, D. R. (Eds.). (2003). *Standards-based school mathematics curriculum: What are they? What do students learn?* Mahwah, NJ: Lawrence Erlbaum.

Synder, J., Bolin, F., & Zumwalt, K. (1992) Curriculum implementation. In P. W. Jackson (Ed.), *The handbook of research on curriculum* (pp. 402-435). New York: MacMillan.

Stein, M. K., Remillard, J. T., & Smith, M. S. (2007). How curriculum influences student learning. In F. K. Lester (Ed.), *Second handbook of research on mathematics teaching and learning* (pp. 319-370). Charlotte, NC: Information Age Publishing.

Stylianides, G. J. (Ed.). (in press). Reasoning-and-proving in mathematics textbooks: From the elementary to the university level. *International Journal of Educational Research* (Special Issue).

Sykes, G. (1990). Organizing policy into practice: Reactions to the cases. *Educational Evaluation and Policy Analysis, 12*(3), 349-353.

Tarr, J. E., Chávez, Ó., Reys, R. E., & Reys, B. J. (2006). From the written to the enacted curricula: The intermediary role of middle school mathematics teachers in shaping students' opportunities to learn. *School Science and Mathematics, 106*(4), 191-201.

Tarr, J. E., Grouws, D. A., Chávez, Ó., & Soria, V. M. (2013). The effects of content organization and curriculum implementation on students' mathematics learning in second-year high school courses. *Journal for Research in Mathematics Education, 44*(4), 683-729.

Tarr, J. E., Reys, R. E., Reys, B. J., Chávez, Ó., Shih, J., & Osterlind, S. J. (2008). The impact of middle-grades mathematics curricula and the classroom learning environment on student achievement. *Journal for Research in Mathematics Education, 39*(3), 247-280.

Thompson, D. R., & Senk, S. L. (2010). Myths about curriculum implementation. In B. Reys, R. Reys, & R. Rubenstein (Eds.), *Mathematics curriculum: Issues, trends, and future directions* (pp. 249-263). Reston, VA: National Council of Teachers of Mathematics.

Thompson, D. R., Senk, S. L., & Johnson, G. J. (2012). Opportunities to learn reasoning and proof in high school mathematics textbooks. *Journal for Research in Mathematics Education, 43*(3), 253-295.

Thompson, D. R., Senk, S. L., & Yu, Y. (2012). *An evaluation of the third edition of the University of Chicago School Mathematics Project Transition Mathematics.* Chicago, IL: University of Chicago School Mathematics Project. Retrieved from http://ucsmp.uchicago.edu/research.

Travers, K. J. (1992). Overview of the longitudinal version of the Second International Mathematics Study. In L. Burstein (Ed.), *The IEA study of mathematics III: Student growth and classroom processes* (pp. 1-14). Oxford, England: Pergamon Press.

Usiskin, Z. (2010). The current state of the school mathematics curriculum. In B. J. Reys, R. E. Reys, & R. Rubenstein (Eds.), *Mathematics curriculum: Issues, trends, and future directions* (pp. 25-39). Reston, VA: National Council of Teachers of Mathematics.

Usiskin, Z. (2013). Studying textbooks in an information age—a United States perspective. *ZDM: The International Journal of Mathematics Education, 45*(5), 713-723.

Valverde, G. A., Bianchi, L. J., Wolfe, R. G., Schmidt, W. H., & Houang, R. T. (2002). *According to the book: Using TIMSS to investigate the translation of policy into practice through the world of textbooks.* Dordrecht, Netherlands: Kluwer.

Venezky, R. L. (1992). Textbooks in school and society. In P. W. Jackson (Ed.), *Handbook of research on curriculum* (pp. 436-460). New York: Macmillan.

Webb, N. L. (2007). Issues related to judging the alignment of curriculum standards and assessments. *Applied Measurement in Education, 20*(7), 7-25.

Wilson, M. R., & Lloyd, G. M. (2000). Sharing mathematical authority with students: The challenge for high school teachers. *Journal of Curriculum and Supervision, 15*(2), 146-169.

Zorin, B. (2011). *Geometric transformations in middle school mathematics textbooks* (Unpublished PhD dissertation). University of South Florida.

CHAPTER 2

EXAMINING VARIATIONS IN ENACTMENT OF A GRADE 7 MATHEMATICS LESSON BY A SINGLE TEACHER

Implications for Future Research on Mathematics Curriculum Enactment

Mary Ann Huntley and Daniel J. Heck

This chapter documents how a seventh-grade mathematics teacher enacted the same *Math Thematics* lesson with two different classes. A brief overview of this textbook series is provided, followed by contextual information about the teacher and detailed information about the specific lesson that was enacted. Vignettes illustrate differences in how students engaged with the mathematical content of the lesson during the two classes. This investigation of enactment serves as a springboard to examine opportunities students are provided to make sense of the mathematical content of a lesson, provides an impetus to develop a conceptual framework to guide research, and suggests ways in which chaos theory might be applied to further research on the enacted mathematics curriculum.

INTRODUCTION

As teachers and as students, we have all experienced the improvisational nature of teaching (Franke, Kazemi, & Battey, 2007; Stein, Engle, Smith, & Hughes, 2006). Even well planned lessons, with rich problems chosen in advance and carefully scripted questions to ask students, play out differently based on a variety of factors. These factors include teacher and student knowledge, beliefs, practices, access to resources, contextual resources, and constraints (Remillard & Heck, this volume). This chapter documents that a seventh-grade mathematics teacher enacted the same lesson in different ways with two different classes. Consistent with Eisenmann and Even (2008), who also report variations in a teacher's use of a single lesson with two sets of students, the data reported here suggest that the two enactments of this lesson potentially led to different opportunities to learn mathematics for the two classes of students. Indeed, the idiosyncrasies and unpredictable aspects of curriculum enactment[1] naturally lead to opportunities for student learning that are distinctly dissimilar (Hiebert & Grouws, 2007).

The goal of the broader research project on which this chapter is based was to investigate enactment of, and develop instruments to measure, implementation of two middle-grades mathematics textbook series: *Connected Mathematics* (Lappan, Fey, Friel, Fitzgerald, & Phillips, 2002) and *Math Thematics* (Billstein & Williamson, 1999-2005). *Math Thematics* is the instructional material used by the teacher who is the focus of this chapter. Data include an audio-taped interview with the lead author of the textbook series (Rick Billstein), and videotaped observations of one teacher's enactment of the same lesson with two different groups of students. During classroom observations, running field notes were taken to supplement the video.

This chapter begins with a brief overview of the *Math Thematics* textbook series, followed by contextual information about the teacher and the specific lesson enacted. The discussion then focuses on differences in how the lesson played out in two class periods, together with analysis of the lesson in terms of the learning opportunities provided to students. The importance of these differences in enactment of the same lesson, in terms of potential student learning, suggests the need for a conceptual framework to guide research on the enacted mathematics curriculum. We conclude the chapter by proposing some ways that, within such a framework, ideas from chaos theory might be applied to research on the enacted mathematics curriculum. That is to say, we consider ways in which potentially important consequences for student learning may arise from seemingly small differences in curriculum enactment.

OVERVIEW OF *MATH THEMATICS*

Math Thematics is a full three-year set of mathematics curriculum materials intended for students in grades 6-8, corresponding roughly to ages 11-14. With funding from the National Science Foundation (NSF), *Math Thematics* was developed to support implementation of the National Council of Teachers of Mathematics' (NCTM) *Curriculum and Evaluation Standards* (NCTM, 1989). In contrast with most other textbooks available at the time, *Math Thematics* places greater emphasis on mathematical investigation, mathematics presented in real-world contexts, connections among content areas of mathematics, connections between mathematics and other disciplines, and integration of technology with mathematics (NSF, 1989).

"Guided discovery" or "guided constructivism with scaffolding" are phrases used by the authors of *Math Thematics* to describe the learning theory that underpins the instructional materials. As described by the lead author, Rick Billstein (personal communication, July 9, 2004), each mathematical concept is introduced with a hands-on activity set in a real-world context. Students are expected to engage in the activities without teacher direction. Then with teacher direction and some repeated practice of ideas, students are expected to abstract and generalize the idea, which students internalize and make their own. The teacher does not just give students algorithms.

Corresponding to each year of *Math Thematics*, there is a hardbound textbook that includes eight modules. Each module has a specific theme that connects the mathematical content to applications in the real world.[2] Each module contains from four to six sections, each further divided into four components: *Setting the Stage*, *Explorations* (from one to three Explorations in each Section), *Summary of Key Concepts*, and *Practice and Application Exercises*. These four components are shown schematically in Figure 2.1.

At the beginning of each section, *Setting the Stage* includes a reading, discussion, or activity through which the teacher creates a context and motivation for students to learn the mathematical content. For this portion of the lesson, intended to be 10-15 minutes in duration, the textbook authors suggest that teachers allow students to solve the *Think About It* questions with minimal guidance; that is, teachers elicit students' ideas and probe for their reasoning. During *Explorations*, students are actively involved in investigating mathematical concepts, learning mathematics skills, and solving problems. Explorations are intended to include a mixture of whole-class, small group, and individual work. Group work typically accounts for about 30% of the time, during which students are actively involved in investigating math concepts, learning math skills, and

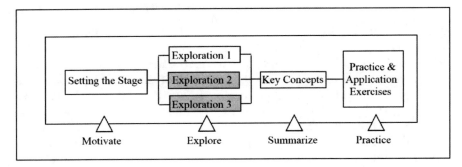

Figure 2.1. Organization of a section of *Math Thematics*.

solving problems, while teachers are encouraged to provide guidance and scaffolding. An important part of each Exploration is formative assessment, which appears in the form of *Checkpoint Problems* that allow the teacher to check students' understanding. Checkpoint Problems provide opportunities for teachers to pause to ensure that each student (or perhaps each group of students) understands the material before moving to new content. The *Key Concepts* summarize ideas to help students study and review the key mathematical ideas of each section, as well as develop note-taking skills. This summary generally occurs after several days of instruction. The *Practice and Application Exercises* give students a chance to practice using the skills and concepts in the Explorations and apply them to solve many types of problems. In addition, at the end of each section the instructional materials provide exercises that can be used to check students' understanding before starting the next section.

CONTEXTUAL INFORMATION

Ms. James[3] teaches seventh-grade mathematics in a public middle school in the Northeast region of the United States. The school serves approximately 600 students in grades 6-8. Approximately 97% of the students are non-Hispanic White, and 20% are eligible for free or reduced-price lunch.[4] When these data were collected, Ms. James had been teaching mathematics for eight years, and this was her third year teaching *Math Thematics* Course 2. During the prior summer, she had attended professional development sponsored by the textbook publisher.

When Ms. James' classroom was visited, she was beginning Module 3 Section 6, titled *Triangles and Equations*. Figure 2.2 contains a description of this section of the textbook.

> In Section 6, students continue their study of the connection between universal languages and mathematics by looking at shapes and symbols that have been used to communicate ideas. The designs of the symbols, many of which are based on geometric shapes, will introduce the lesson on triangle construction in this section. Students will learn definitions for isosceles, equilateral, and scalene triangles, and they will use a straightedge and a compass to construct triangles. As they learn how to use a compass to construct a triangle, they will be introduced to chords, radii, diameters, and arcs of a circle. Students will use their triangle constructions to examine the triangle inequality theorem informally. Their explorations with shapes will lead students to investigate how the dimensions in a flag are related. They will write verbal expressions to describe the relationship, and then will translate the verbal expression into equivalent multiplication and division equations. Students will use inverse operations to solve the equations. They will also check their solutions.

Source: Buck et al. (2002, p. 3-47).

Figure 2.2. Description of the lesson: Course 2 Module 3 Section 6.

The foci of the observations reported in this chapter, henceforth referred to as "the lesson," were the *Setting the Stage* and *Exploration 1* portions of Grade 7 Module 3 Section 6. The corresponding pages from the teacher's edition of the book are provided in Appendix B.[5] The teacher used the Labsheet that accompanies the lesson, which is a supplemental material provided by the publisher (see Appendix C[6]).

The mathematical objectives for the lesson as stated by the authors are:

- Classify triangles by side lengths
- Use a ruler and a compass to construct triangles with given side lengths
- Determine which side lengths will form a triangle. (Buck et al. 2002, p. 3-47)

Two different classes were observed when Ms. James was teaching this lesson. First, Ms. James taught the lesson to her Period 1 class, which had 18 students. Two days later she taught the lesson to her Period 2 class, which had 21 students. Both observations occurred during a split class period, 10:55 A.M.-12:20 P.M., which includes a 20-minute break for lunch.[7] There were two aides present for the Period 2 class, and none during the Period 1 class. The aides assisted with homework review and provided support to students during small-group work.

LESSON ENACTMENT

During classroom observations in both Periods 1 and 2, Ms. James followed the *Math Thematics* textbook quite closely, with the textbook as the only source of content. The Labsheet that accompanies this lesson was distributed to students to use in answering questions 3-5. During the Exploration, there was a mixture of whole-class, small group, and individual work. Throughout the lesson there was some direct instruction. Students did each of the Checkpoint questions, and Ms. James discussed the answers to each question in a whole-class discussion. At the same time, there were several notable variations in enactment of this lesson across the two classes. These differences are detailed in the sections that follow.

Use of Instructional Time

Each class had 65 minutes of instructional time available, yet this time was not used in the same manner in the two classes. Specifically, the beginning of class played out differently across Periods 1 and 2, as shown in Figure 2.3 and discussed below.

In Period 1, Ms. James began by checking students' homework while two of the students wrote answers to homework problems on the board. These were Checkpoint problems from an earlier lesson. Ms. James then quickly read the answers to other assigned homework problems. Attention to new material for the lesson began 13 minutes into the period, with students working in pairs using a "Notes" page that Ms. James had developed for students to write definitions to key terms listed in the right-hand column on page 217 of the textbook (see Appendix B). Ms. James then led a whole-class discussion of the terms and definitions, writing the definitions on a Notes page projected in front of the class. These activities

Period 1	Homework Review (12 min)	Define New Terms (13 min)	Setting the Stage Exploration (40 min total, 22 min in group work)
Period 2	Warm Up (7 min)	Define New Terms (9 min)	Setting the Stage Exploration (49 min total, 28 min in group work)

Figure 2.3. Sequence of instructional activities as enacted in Periods 1 and 2.

accounted for the first 25 minutes of the class period, at which time the Setting the Stage portion of the lesson began.

In Period 2, class began with the aides checking students' homework. Ms. James collected another homework assignment and told students to do the following warm-up problems:

1. Name the operation that will undo division.
2. Define scalene.
3. Will 3.2 ft, 3.4 ft, 4.2 ft make a triangle? If so, name it.

Four minutes into the class instructional time, Ms. James led a whole-class discussion of these warm-up problems. Answers were given to the first two problems, but not the third. For the third, Ms. James probed some students' thinking. One student hypothesized that the given side lengths will not form a triangle because all three lengths have to be close together. Another student conjectured that a triangle could be constructed, and that the triangle would be a scalene triangle. Another student offered the idea that "you add them all up, or something like that." Ms. James concluded this discussion by saying, "We're actually going to do this today. So I'm not going to reveal how to do that yet. I'm going to leave a question mark and we'll come back to see if that will or will not make a triangle by the end of class .... I was just seeing kind of who had an idea and who didn't."

These warm-up problems, and the brief exchange and lack of resolution regarding the third problem in particular, provided a means for students to consider their prior knowledge about triangles. The fact that this activity occurred in one class period and not the other seems particularly significant in light of current learning theory that suggests the importance of activating prior knowledge and motivating a need for learning new information (Bransford, Brown, & Cocking, 2000).

After collecting a third homework assignment, Ms. James began the Period 2 lesson eight minutes into the class instructional time, and at this point followed the same format as she did in Period 1. Students worked in pairs to write definitions to the key terms listed in the right-hand column on page 217 of the textbook, and then Ms. James led a whole-class discussion of the terms and definitions. Ms. James concluded this activity 16 minutes into the class period, and then began the Setting the Stage portion of the lesson.

In each class, the remainder of time was spent on Setting the Stage and Exploration 1. The available time for these parts of the lesson was of different duration for each class. In Period 1, students devoted 40 minutes to these portions of the lesson, in contrast to 49 minutes by students in Period 2. The nine-minute difference between the start-up times in the two classes accounted for roughly 14% of the available instructional time.

The difference in how instructional time was used at the beginning of each class period potentially affected student learning, as students in Period 2 had more time to work through the core problems of the lesson.

Engagement with the Mathematical Content

Differences in how the beginning of each class period played out may have affected how students ultimately engaged with the mathematical content of the lesson. In terms of time, group work accounted for approximately 33% of class time in Period 1, and 43% of class time in Period 2. Moreover, the fact that Period 1 had less instructional time for the lesson may have been the reason Ms. James skipped one particular question (question 9) during that class. The potential consequence of this decision is highlighted by considering the instructional sequence surrounding question 9.

Question 8 involves students constructing a triangle with side lengths 8 cm, 6 cm, and 4 cm. Students use a ruler and compass for this construction. Question 9 then asks students to construct two triangles with given side lengths. The first construction is not possible (side lengths 1 cm, 3 cm, 6 cm), but the second is possible (side lengths 2 cm, 3 cm, 4 cm). Question 9, in turn, scaffolds question 10a, which asks, "What relationship must exist among the lengths of three segments for them to form a triangle?" According to the Teacher's Resource Book (Buck et al., 2002), "Using a compass and experimenting with constructions using different side lengths, one discovers that the sum of the length of two sides of a triangle must be greater than the length of its remaining side." The opportunity for students to make this discovery in answering question 10a, current learning theory would suggest (Bransford et al., 2000), grows out of their intellectual engagement in the two scenarios presented in question 9. Not providing students with the opportunity to form triangles with various side lengths short-circuits this instructional sequence. Thus, students who do not engage with forming a triangle with side lengths that are not possible are led either to rely on learning in a prior math course or to guess at the relationship called for in question 10a. This shortcoming, illustrated in Vignette 1, is indeed what appeared to arise in Ms. James' Period 1 class as students discussed question 10a.

Vignette 1

Ms. James: So it's easy to construct a triangle if you've got the proper tools .... We have to decide if we can make triangles if we don't have the tools we need to construct them .... OK? So, let's look at the relationship of the lengths of the sides.

	Look at these sides right here – 8, 6, and 4. Can someone come up with a rule to tell me, because we know that an 8 cm, 6 cm, 4 cm *will* make a triangle? Can someone come up with a rule that might always work ... whether a triangle can be constructed or not? Just looking at those dimensions?
Paul:	[inaudible]
Ms. James:	Well, you're right. All of the angle measures have to add up to 180, but we're not looking at angle measures today; we're looking at side lengths, so that rule doesn't work. Tony, what was your idea?
Tony:	Uhm ... doesn't the bottom have to be either the same or larger than the other two sides? ....
Ms. James:	Tony had an idea. Refresh your memory. He said that the longest side, or the base, but sometimes people think the base is the bottom ... he said the longest side, these other two measurements had to be the same or longer than the longest side. So I said, let's investigate that. So we did [a construction] where we changed this [side from 6 cm] to 4 cm. So then if this is 4 cm ... and this is 4 cm, if I put these together, that's the same distance as 8 cm, correct? Could I make a triangle? [pause] No. So I have to come up with a different rule. What do you think the rule might be? Because we're ... on the right track. We're on the path to that rule. It can't be the same ... we just proved it can't be the same.
Tony:	Longer.

Ms. James does not acknowledge Tony's response and calls on Charles, whose hand is raised.

Charles:	What if you have 8 cm ... and then both the arcs are 8 cm ... [inaudible]
Ms. James:	If they are both 8 cm? And then 8 and 8—what's that?
Charles:	16.
Ms. James:	16. So then that's bigger than the other one, right? So then it would probably work. Do you want me to try it?

Ms. James constructs another triangle, using side lengths 8 cm, 8 cm, and 8 cm.

Ms. James:	So the two other sides can be the same, but what's the difference? .... The last time I had 4 and 4 and here I have 8 and 8.
Dan:	Never mind.

Ms. James:	Never mind?
Sue:	[inaudible]
Ms. James:	They can't add up to this, but what does it have to add up to?
Tony:	Double the number?
Ms. James:	Not double, although in this case it was double.
Kay:	180?
Ms. James:	Nope. I'm not looking at degrees. I'm looking at sides.
Anna:	24?
Ms. James:	They don't have to add up to 24. That might be a pattern. These two sides have to add up to be something what, compared to this?
Tony:	Greater.
Ms. James:	The two sides added together have to be greater than the third side.

In contrast to Period 1, after completing question 8 as a whole class (i.e., learning how to use a ruler and compass to construct a triangle with given side lengths), students in Period 2 were told to work on question 9 with their partners (i.e., students were to try constructing a triangle with side lengths 1 cm, 3 cm, 6 cm, and another with side lengths 2 cm, 3 cm, 4 cm). While students worked on this question, Ms. James monitored students' progress and answered their questions. Noticing that students were off task, Ms. James modeled the (impossible) construction in question 9a. Then she continued monitoring, and several minutes later she completed the construction in question 9b on the overhead projector (silently, as students continued working with their partners). The subsequent classroom discourse related to question 10 is outlined in Vignette 2.

Vignette 2

Ms. James:	So, we have three constructions. We were able to make one [a triangle] in number 8. We discovered, when we started together [as a whole class], that, can I make a triangle here?

Ms. James points to her attempt to draw a triangle with side lengths 6 cm, 1 cm, 3 cm, as shown below.

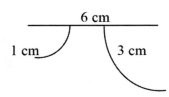

Students: No.
Ms. James: Why can't I make a triangle here? What's missing? What happened when I started to construct it? Fred?
Fred: They didn't overlap.
Ms. James: The arcs did not intersect. So this means there was no point where they overlapped each other, as Fred said. And in the last one, were you able to make a triangle? [Slight pause.] You should have been able to make a triangle, because they would overlap. I've written down the measurements that we used in each, so we need to try to come up with a rule for what will make a triangle and what will not make a triangle. So who has an idea that we can either investigate, or maybe your idea is right on the money. Bob?
Bob: Uhm, could you add both of ... the two smallest numbers together ... and if they're larger than the biggest number, then you can make a triangle.
Ms. James: So you're saying if I add two sides, and sum ...
Bob: Add two smaller sides.
Ms. James: Okay. You're saying smaller. And their sum, what you get is ...
Bob: Bigger than the biggest number.
Ms. James: Greatest than largest number, you get a triangle.

While Ms. James is saying this, she writes Bob's conjecture on the overhead projector.

Ms. James: So let's test out Bob's idea. Who can give me three measurements, where two sides, two smaller sides, when you add them, that would be larger, greater, than the third side?
Chris: [inaudible]

Ms. James constructs a triangle with side lengths 3 cm, 3 cm, 4 cm.

Ms. James: Will it make a triangle?
Students: Yes.
Ms. James: So Bob, I think you're on the right track! Your idea works! Let's test it in 8 and 9.

Ms. James leads a whole-class discussion, implementing Bob's conjecture on the sets of side lengths given in problems 8 and 9.

Ms. James: So it looks like it works! We've done it many times and we haven't found a case where it hasn't worked. So your task

	right now is: While I collect my compasses and rulers, is to do the Checkpoint, number 11 and show me your paper when you're done.
Bob:	Was I right?
Ms. James:	You were right, Bob! Very good job!
Bob:	I just guessed!
Ms. James:	Well, you must have had some kind of inkling, to lead you that way.

DISCUSSION

We have illustrated the enactment of a mathematics lesson for two different classes taught by the same teacher. These serve as a springboard to examine opportunities students are provided to make sense of the mathematical content of a lesson, provide the impetus for developing a conceptual framework to guide research, and suggest ways in which chaos theory might be applied to research on the enacted mathematics curriculum.

Providing Opportunities for Sense Making

The dialogue portrayed in the vignette from Period 1 shows Ms. James endeavoring to conclude the lesson with appropriate sense making about the third learning goal of the lesson, *Determine which side lengths will form a triangle*. This intent is particularly evident in her statement, "We have to decide if we can make triangles if we don't have the tools we need to construct them." It is notable that mathematics lessons in the U.S. have historically shown little evidence of conceptual sense making of this sort (Hiebert et al., 2003; Weiss, Pasley, Smith, Banilower, & Heck, 2003), so finding the class working at making sense of this learning goal appeared promising. In this case, however, the effort at sense making did not appear effective for the students—Ms. James was the one doing most of the intellectual work, while the students seemed to be suggesting a variety of ideas they could recall related to triangles. Why might this be?

The answer may lie in what occurred, or did not occur, earlier in the lesson. Recall that the students in Period 1 did not experience the warm-up problems that might have activated important prior knowledge about triangles and motivated a need to learn the triangle inequality relationship. Also, the instructional sequence in *Math Thematics* is designed for students to do more constructions with combinations of side lengths that make it either possible or impossible to form triangles. These explorations provide students opportunities for intellectual engagement that could guide conjecturing, testing of ideas, and discovery. Unfortunately, the teacher was not interviewed after either enactment of this lesson, so we have no insight as to why she handled the lesson differently in the two classes. However, we argue that if the students had experienced opportunities like those in question 9, which are pivotal in addressing the third learning goal of the lesson, they may not have struggled so mightily with the relationship desired in question 10a. In fact, a student in Period 2, having completed the third warm-up question that was likely to have activated prior knowledge, together with the opportunity to work on the explorations in question 9, readily described the triangle inequality relationship (question 10a).

We are not arguing that in *any* enactment of this lesson in which these two activities take place (the third warm-up question and question 9 from the textbook), the same result will occur (correct answer to question 10a). However, we do suggest that the *likelihood* of achieving the desired outcome is higher when these activities are completed than if they are omitted. To clarify, we have no information about actual differences in student learning that may have occurred with Period 1 versus Period 2 students, as this was beyond the scope of the study. Here we are arguing about *potential* differences in student learning as a result of the different enactments of the same lesson.

The Need for a Conceptual Framework

The November 2010 Conference on Research on the Enacted Mathematics Curriculum, held in Tampa, Florida, began with participants watching a videotape of a portion of the enactment of this *Math Thematics* lesson from Ms. James's Period 1 classroom. Conference participants were given the instructions provided in Figure 2.4.

After viewing the videotape and being given time for writing individual thoughts, participants, who were seated in small groups, briefly shared what they wrote about their observations, paying attention to differences in what people attended to and/or different terms used to describe the enacted curriculum in the videotaped lesson. They created a written document to capture how observations, perspectives, or language (i.e., use of terms or definitions) differed across the people at their table.

Many issues and challenges arose when conference participants discussed this enactment of the lesson, including the following:

> Imagine you have been given the assignment to observe/study the enacted mathematics curriculum in seventh-grade classrooms in a district. You decide to take field notes to capture your observations. In a few minutes you will watch a short video segment from an actual seventh-grade mathematics classroom. The teacher is using a lesson from the Math Thematics curriculum, and on the day you observe, she is covering the first lesson of Module 3 Section 6. This section of the book focuses on triangles and equations, and the lesson you will observe is about constructing triangles. Please take two minutes to read the lesson, using the handouts provided in your registration folder. The lesson is called "Symbols of the People" on pages 216-219, with accompanying Labsheet 6A. Remember that when you watch the video you should take field notes, focusing on the "enacted curriculum." After viewing the video segment and taking notes, you will have a few minutes to review and analyze what you wrote so that you are ready to talk about your observations with other people at your table.

Figure 2.4. Instructions for video task given to conference participants.

- One person pondered whether the lesson was of high fidelity to the written materials, and how high fidelity might differ from high quality enactment.
- One participant questioned the difference between fidelity to the written lesson versus fidelity to the mathematical content of the lesson.
- Other conference participants discussed the concept of fidelity of implementation but used different words, for instance: faithful implementation, matching the written curriculum, coverage of material, and spirit of the curriculum.
- One person's written notes focused on the discourse throughout the lesson (e.g., the use of particular words that denote power and authority).
- Another person focused on task enactment, including teacher versus student activities and dialogue throughout the lesson.
- Other ideas that were discussed included:
  o Teacher decision making about what was skipped and why;
  o Mathematics content considerations (e.g., the mathematical storyline of the lesson, extent of justification or proof offered during the lesson);
  o Pedagogical considerations and decision-making (e.g., teacher questioning, wait time);
  o Relationships among teacher beliefs, teacher knowledge, and teacher understanding;

- Influence of the room arrangement on the lesson enactment; and
- The school and district context as an influence on the lesson enactment.

This activity made clear that the phrase *enacted curriculum* means different things to different researchers, or at least prompts them to pay attention to different aspects of what takes place in a mathematics classroom. There was explicit discussion of aspects in the Period 1 lesson that our analysis suggests might be critical to students' opportunities to learn mathematical content. The diversity of perspectives offered by conference participants reflected their experiences and interests with respect to the conference theme—research on the enacted mathematics curriculum. This opening activity provided the impetus for developing a conceptual framework, not to narrow or restrict research on the enacted curriculum, but rather, to give the field a shared language to guide research design and to allow for comparing and combining studies to accumulate knowledge and generate new hypotheses. The resulting conceptual framework is the subject of the chapter in this volume by Remillard and Heck.

Implications for Research

Interactions between teachers and students in classrooms are not deterministic systems. We cannot say with any certainty that if a teacher does X, then a student will do Y. Quite the contrary is true. Small changes in the enactment of instructional materials, such as subtle differences in dialogue, small changes to the pacing, modifications to the instructional sequence, and variations in classroom interactions between the teacher and students, can lead to vastly different opportunities for student learning. Consequently, we suggest that research on the enactment of the school mathematics curriculum might benefit by application of chaos theory, which is predicated on the idea that small perturbations to the initial conditions of a system can lead to vastly different outcomes (Gleick, 1987). For instance, as in meteorology, in which complex mathematical models are run repeatedly on high-speed computers with slightly different initial conditions in an effort to understand the variability in weather forecasts, researchers studying the enacted curriculum might conduct more systematic studies of how small changes in dialogue, pacing, instructional sequence, the use of tools, or classroom interactions affect opportunities for students to achieve specific learning goals. As another example, in research on the enacted curriculum, we wonder whether there is an analog to the Lyapunov exponent, which characterizes the extent of sensitivity of a chaotic system to initial conditions. Applying chaos theory, together with what we know about how students learn, to the study of enactment of mathematics instructional materials might be a fruitful avenue for further research to understand what contextual conditions of enactment and factors in enactment influence student learning opportunities, how this influence plays out, for whom, and with what consequences.

APPENDIX A

Math Thematics Module Titles and Mathematical Content

Table A1. Math Thematics Book 1

Module	Title	Mathematical Content
1	Tools for Success	Patterns and sequences; lines, angles, and triangles; a problem solving approach; estimation, mental math, or a calculator; using visuals; problem solving skills
2	Patterns and Designs	Polygons and line symmetry; fractions and mixed numbers; equivalent fractions; transformations; understanding decimals; decimal addition and subtraction
3	Statistical Safari	Sets and metric measurement; fractions and percents; bar graphs and line plots; mean, median, mode; dividing decimals, estimation, and mental math; stem-and-leaf plots and dividing by a decimal
4	Mind Games	Probability; factors and divisibility; fraction multiplication; decimal multiplication; equations and graphs; multiples and mixed numbers
5	Creating Things	Comparing fractions; customary units of length; addition and subtraction of fractions; addition and subtraction of mixed numbers; capacity and mixed number multiplication; division with fractions
6	Comparisons and Predictions	Exploring ratios; rates; using ratios; proportions; geometry and proportions; percents and probability
7	Wonders of the World	Area; space figures; weight in the customary system; circles and circumference; circles and cylinders; temperature, integers, and coordinate graphs
8	Our Environment	Adding and subtracting integers; line graphs, scientific notation, and percent; metric capacity and percent; geometric probability; misleading graphs and averages

(Appendix A continues on next page)

APPENDIX A (CONT.)

Math Thematics Module Titles and Mathematical Content

Table A2. Math Thematics Book 2

Module	Title	Mathematical Content
1	Making Choices	Data displays; sequences and exponents; probability; problem solving; assessing problem solving; expressions and representations
2	Search and Rescue	Looking at angles; integers and coordinates; integer addition and subtraction; function models; addition and subtraction equations
3	A Universal Language	Factors, divisibility, and multiples; fractions and tree diagrams; fractions and mixed numbers; decimals and exponents; metric units of length; triangles and equations
4	The Art of Motion	Circumference; fraction multiplication and division; decimal multiplication and division; rotations and reflections; multiplication and division of integers; translations, similarity, and two-step equations
5	Recreation	Ratios and data displays; proportions and plots; percent; percent and probability
6	Flights of Fancy	Inequalities, polygons, and probability; square roots, surface area, and area of a circle; triangles and similarity; parallel lines and angles in polygons; volume of a prism, and metric relationships
7	Health and Fitness	Cylinders and graphs; percent equations; customary capacity and inequalities; circle graphs, and choosing a graph; quadrilaterals
8	Heart of the City	Counting problems; drawing views and finding volumes; permutations and combinations; tessellations and volumes of cones

(Appendix A continues on next page)

APPENDIX A (CONT.)

Math Thematics Module Titles and Mathematical Content

Table A3. Math Thematics Book 3

Module	Title	Mathematical Content
1	Amazing Feats, Facts, and Fictions	Problem solving and rates; displaying data; scatter plots; circumference and volume; equations and expressions; area and perimeter
2	At the Mall	Proportions and percents; working with percents; exploring probability; operations with integers; operations with fractions; inequalities and box-and-whisker plots
3	The Mystery of Blacktail Canyon	Square roots and measurement; equations and graphs; slope and equations; similar figures and constructions; scientific notation and decimal equations; logical thinking
4	Patterns and Discoveries	Fractals, sequences, and triangles; rotation and numbers; equations with fractions and quadrilaterals; rational numbers and polygons; working with triangles; geometry and probability
5	Inventions	Working with cylinders; slope and equations of lines; working with exponents; complements, supplements, and tangents; counting techniques; working with probability
6	Architects and Engineers	Geometry and perspective; geometry constructions; surface area and volume; angles formed by intersecting lines; solving inequalities; scale drawings and similar figures
7	Visualizing Change	Graphs and functions; linear equations and problem solving; modeling exponential change; algorithms and transformations; exploring quadratic functions
8	Making an Impact	Collecting data; making data displays; representing data; equivalent rates and relative frequency

Examining Variations 39

APPENDIX B

Module 3 Section 6 Triangles and Equations

Section 6 Triangles and Equations

IN THIS SECTION
EXPLORATION 1
• Constructing Triangles

EXPLORATION 2
• Multiplication and Division Equations

WARM-UP EXERCISES

State the operation that undoes the given operation.
1. addition
   subtraction
2. subtraction
   addition
3. multiplication
   division
4. division
   multiplication
5. State two properties of triangles.
   Sample Response: They have 3 sides; They have 3 angles; The sum of their angle measures is 180°.

1. a. Sample Response: a square on the checkered flag
b. one of the halves formed by the diagonal in the first flag
c. the angles formed by the diagonal of the navigation flag
d. Sample Response: one of the corners of the middle flag
e. the two triangles formed by the diagonal in the first flag

2. a. Sample Response: to signal a car has crossed the finish line in a car race

···Setting the Stage

People have been sending messages using shapes and symbols for hundreds of years. Flags and banners can express ideas and communicate information without words. The designs of many of these flags are based on geometric concepts.

Navigation	Humanitarian Services	Sports
This flag shows the *International Code of Signals* symbol for "person overboard."	In many countries of the world this *Flag of Healing* is the sign of emergency medical assistance.	Flags such as this checkered flag are used in some sports to signal the referee or the participants.

Think About It

1 Find an example of each geometric figure on one of the flags.
   a. a rectangle   b. a triangle   c. an acute angle   d. a right angle
   e. a pair of triangles that are the same size and shape

2 a. What is the checkered flag used for?
   b. What fraction of the checkered flag is black? $\frac{18}{35}$

216 Module 3   A Universal Language

(Appendix B continues on next page)

APPENDIX B (CONT.)

Module 3 Section 6 Triangles and Equations

Exploration 1

Constructing TRI△NGLES

SET UP You will need: • Labsheet 6A • metric ruler • compass

Many nations around the world have a national flag to symbolize their country. Some of these flags contain triangles.

Use Labsheet 6A for Questions 3–5.

3 Follow the directions on the labsheet to measure and group the triangles on the *Flags with Triangles*. See Additional Answers.

4 **Discussion** One way to classify triangles is by the lengths of their sides. Two segments that are equal in length are congruent. Did anyone in your class group the triangles using 0, 2, and 3 congruent sides? What other groupings were used?
Sample Response: Yes; horizontal stripes, diagonal stripes, no stripes

Some of the triangles you measured on Labsheet 6A had at least two congruent sides. These triangles are isosceles. Triangles with three congruent sides are equilateral. If all its sides are different lengths, a triangle is scalene.

5 **a.** Classify each triangle you measured as *isosceles*, *equilateral*, or *scalene*.

   b. Which triangles can be classified more than one way? Why?

6 **✓ CHECKPOINT** Classify the triangle with each set of side lengths as *isosceles*, *equilateral*, or *scalene*. Be as specific as possible.
   a. 7 in., 5 in., 3 in.   **b.** 1 m, 1 m, 1 m   **c.** 2 cm, 4 cm, 4 cm
       scalene   equilateral and isosceles   isosceles

7 Use your ruler to draw each triangle and label the side lengths.
   a. isosceles triangle that is not equilateral   **b.** equilateral triangle
   c. scalene triangle with 5 cm, 8 cm, and 9 cm sides
   Check students' drawings.

Some triangles are more difficult to draw than others. A compass is one tool that can make the job easier if you know how to use it.

GOAL

LEARN HOW TO...
* classify triangles by side length
* construct triangles and circles
* determine which side lengths will form a triangle

AS YOU...
* examine flags

KEY TERMS
* congruent
* isosceles
* equilateral
* scalene
* circle
* center
* radius
* chord
* diameter
* arc
* construction

5. a. isosceles triangles: Philippines, Bahamas, Sudan, Equatorial Guinea; equilateral triangles: none; scalene triangles: Tanzania, Trinidad and Tobago
b. equilateral triangles; Triangles with 3 congruent sides are equilateral. Since they contain at least 2 congruent sides, they are also isosceles.

✓ QUESTION 6
...checks that you can classify triangles.

7. a.

b.

(Appendix B continues on next page)

Examining Variations 41

APPENDIX B (CONT.)

Module 3 Section 6 Triangles and Equations

Student Resource

Compasses and Circles

You can use a compass to construct a *circle*.

To construct a circle, mark the *center* C of the circle. Place the tip of the compass on C and make a complete rotation.

A *circle* is the set of all points in a plane that are the same distance from a given point called the **center**.

A segment such as $\overline{CD}$ whose endpoints are the center and any point on the circle is a **radius** of the circle. The length of any radius is called *the* radius.

A segment such as $\overline{AB}$ whose endpoints are both on the circle is a **chord**.

Any chord such as $\overline{EF}$ that passes through the center of a circle is a **diameter** of the circle. The length of any diameter is called *the* diameter.

8 Construct a triangle with sides of length 4 cm, 6 cm, and 8 cm.
Check students' drawings.

An *arc* is part of a circle.

Step 1 Draw an 8 cm segment with endpoints A and B.

Step 2 Open your compass to a radius of 6 cm. Place the tip on A. Draw an *arc* below $\overline{AB}$.

Step 3 Open your compass to a radius of 4 cm. Place the tip on B. Draw an arc that intersects the first arc.

Step 4 Label the intersection point C. Draw segments $\overline{AC}$ and $\overline{BC}$ to complete the triangle.

Module 3 A Universal Language

(Appendix B continues on next page)

APPENDIX B (CONT.)

Module 3 Section 6 Triangles and Equations

You can draw some figures using just a straightedge and a compass. Drawings made this way are called **constructions**.

9 If possible, construct a triangle with the side lengths given. If not, explain why not. Label the sides with their lengths.

    a. 1 cm, 3 cm, 6 cm   **b.** 2 cm, 3 cm, 4 cm

9. a. not possible; It is impossible to make the sides intersect.
b. possible; Check students' drawings.

10 a. Try This as a Class What relationship must exist among the lengths of three segments for them to form a triangle?
The sum of the lengths of any 2 sides must be greater than the third side.
    b. Change one side length in a set of three side lengths in Question 9 so that they will form a triangle.
Sample Response: Change the 1 cm side to 4 cm.

11 ✓ **CHECKPOINT** Tell whether segments with the given lengths *can* or *cannot* form a triangle. If they can, construct the triangle.

    a. 2 cm, 3 cm, 8 cm   **b.** 5 cm, 7 cm, 9 cm
       cannot   can; Check students' drawings.

✓ **QUESTION 11**
...checks that you can tell whether 3 segments form a triangle, and that you can construct a triangle.

HOMEWORK EXERCISES See Exs. 1–10 on p. 224.

Exploration 2

Multiplication and Division Equations

SET UP You will need a metric ruler.

GOAL

LEARN HOW TO...
* write and solve multiplication and division equations

AS YOU...
* explore size relationships in flag designs

A country's flag may come in different sizes, but the relationships between the shapes in the design must stay the same.

12 The state flag of Spain is shown.

    a. Measure the widths of the stripes to the nearest centimeter.

    b. Describe the relationship between the widths.

12. a. 1 cm, 2 cm, 1 cm
b. Possible answers: The width of the yellow stripe is twice the width of a red stripe. The width of a red stripe is $\frac{1}{2}$ the width of the yellow stripe.

Examining Variations 43

APPENDIX C

Lab Sheet 6a: Flags With Triangles

Name _____ Date _____

MODULE 3   **LABSHEET 6A**

Flags with Triangles (Use with Questions 3–5 on page 217.)

Directions
- Measure the lengths of the sides of one triangle on each flag to the nearest millimeter. Record the measurements next to each flag.
- Based on the lengths of the sides, separate the triangles you measured into 3 groups and complete the first two blank columns of the table.
- In the last column of the table, explain your method for grouping the triangles.

Philippines

Sudan

Tanzania

Equatorial Guinea

Bahamas

Trinidad and Tobago

Group	Countries	Lengths of sides of triangles (mm)	Reason for grouping
1			
2			
3			

3-62 Math Thematics, Book 2   Copyright © by McDougal Littell Inc. All rights reserved.

ACKNOWLEDGEMENT

Funding for data collection was provided by a National Academy of Education/Spencer Postdoctoral Fellowship Award to the first author.

NOTES

1. By "lesson enactment," we mean the interactions between teachers and students around the tasks of a lesson.
2. Appendix A lists the modules in the textbook series and the mathematical focus of each module.
3. All names in this chapter are pseudonyms.
4. Free or reduced price lunches are offered to students meeting certain federal guidelines. This statistic is provided here as an indicator of the percentage of students coming from low socioeconomic status families.
5. From *MATH THEMATICS*, Teacher's Edition, by Rick Billstein and Jim Williamson, et al. Copyright ©1999 by McDougal Littell Inc. All rights reserved. Reprinted by permission of Houghton Mifflin Harcourt Publishing Company.
6. From *MATH THEMATICS*, Book 2, Resource Book Modules 3 & 4. Copyright ©1999 by McDougal Littell Inc. All rights reserved. Reprinted by permission of Houghton Mifflin Harcourt Publishing Company.
7. This school uses a rotating class schedule.

REFERENCES

Billstein, R., & Williamson, J. (1999-2005). *Middle grades Math Thematics.* [*student and teacher books 1-3*]. Evanston, IL: McDougal Littell.

Bransford, J. D., Brown, A. L., & Cocking, R. R. (Eds.). (2000). *How people learn: Brain, mind, experience, and school*. Washington, DC: National Academy Press.

Buck, M., Denny, R., Howard, J., Morse, S., Runkel, P., Sanders-Garrett, T., & Tuckerman, C. (2002). *Math Thematics teacher's resource book* [Modules 3 and 4]. Evanston, IL: McDougal Littell.

Eisenmann, T., & Even, R. (2008). Similarities and differences in the types of algebraic activities in two classes taught by the same teacher. In J. T. Remillard, B. A. Herbel-Eisenmann, & G. M. Lloyd (Eds.), *Teachers' use of mathematics curriculum materials: Research perspectives on relationships between teachers and curriculum* (pp. 152-170). New York, NY: Routledge.

Franke, M., Kazemi, E., & Battey, D. (2007). Mathematics teaching and classroom practice. In F. K. Lester (Ed.), *Second handbook of research on mathematics teaching and learning* (pp. 225-256). Charlotte, NC: Information Age Publishing.

Gleick, J. (1987). *Chaos: Making a new science*. London, England: Cardinal.

Hiebert, J., Gallimore, R., Garnier, H., Givvin, K. B., Hollingsworth, H., Jacobs, J., et al. (2003). *Teaching mathematics in seven countries: Results from the TIMSS*

1999 video study (NCES 2003-013 Revised). Washington, DC: U.S. Department of Education.

Hiebert, J., & Grouws, D. A. (2007). The effects of classroom mathematics teaching on students' learning. In F. K. Lester (Ed.), *Second handbook of research on mathematics teaching and learning* (pp. 371-404). Charlotte, NC: Information Age Publishing.

Lappan, G., Fey, J., Friel, S., Fitzgerald, W., & Phillips, E. (2002). *The Connected Mathematics Project. [student and teacher texts]*. Glenview, IL: Prentice Hall.

National Council of Teachers of Mathematics. (1989). *Curriculum and evaluation standards for school mathematics*. Reston, VA: Author.

National Science Foundation. (1989). *Materials for middle school mathematics instruction: Program solicitation*. Washington, DC: Author.

Remillard, J. T., & Heck, D. J. (2014). Conceptualizing the enacted curriculum in mathematics education. In D. R. Thompson & Z. Usiskin (Eds.), *Enacted mathematics curriculum: A conceptual framework and research needs* (pp. 121-148). Charlotte, NC: Information Age Publishing.

Stein, M. K., Engle, R. A., Smith, M. S., & Hughes, E. K. (2006). Orchestrating productive mathematical discussions: Helping teachers learn to better incorporate student thinking. *Mathematical Thinking and Learning, 10*, 313-340.

Weiss, I. R., Pasley, J. D., Smith, P. S., Banilower, E. R., & Heck, D. J. (2003). *Looking inside the classroom: A study of K-12 mathematics and science education in the United States*. Chapel Hill, NC: Horizon Research.

CHAPTER 3

INFLUENCE OF MATHEMATICS CURRICULUM ENACTMENT ON STUDENT ACHIEVEMENT

Patricia D. Hunsader and Denisse R. Thompson

Opportunity to learn has been shown to be a major, if not the most significant, predictor of student achievement in mathematics. Although opportunity to learn is certainly related to the written curriculum to which students are exposed, it is also influenced by how that curriculum is enacted in the classroom. In this chapter, we discuss components of classroom implementation that are often measured in studies of curriculum and achievement. We then review a number of studies in which student achievement was considered in relation to the manner in which the curriculum was implemented. We discuss some of the challenges in researching the implementation and achievement connection and raise questions for future research.

INTRODUCTION

The mathematics curriculum, as embodied in textbooks, plays a major role in determining opportunities students have to learn mathematics

(Grouws & Smith, 2000; Stein, Remillard, & Smith, 2007; Weiss, Pasley, Smith, Banilower, & Heck, 2003). As noted by Ball and Cohen (1996), curriculum materials "are the stuff of lessons and units, of what teachers and students do" (p. 6). Although the textbook, as the *potentially implemented curriculum*, is not itself the curriculum, the textbook does provide an organizational structure for the curriculum and for teachers as they plan for and organize their classroom instruction. "The curriculum is also how a teacher interprets or uses such texts" (Love & Pimm, 1996, p. 398). The textbook, then, serves as a mediator between the intentions, expectations, and goals of curriculum developers and the instruction that occurs in the classroom (Valverde, Bianchi, Wolfe, Schmidt, & Houang, 2002).

Opportunity to learn has long been used by curriculum researchers as a means to interpret achievement results in various international studies (Husen, 1967; Schmidt, Wolfe, & Kifer, 1992). Curriculum researchers have also used opportunity to learn in curriculum evaluation or effectiveness studies to understand achievement differences when students study from different curricula (e.g., Cai, Wang, Moyer, Wang, & Nie, 2011; Grouws, Tarr, Chávez, Sears, Soria, & Taylan, 2013; Tarr, Grouws, Chávez, & Soria, 2013; Tarr, Reys, Reys, Chávez, Shih, & Osterlind, 2008; Thompson, Senk, Witonsky, Usiskin, & Kaeley, 2001; Thompson, Witonsky, Senk, Usiskin, & Kaeley, 2003; Thompson, Senk, & Yu, 2012). Indeed, "opportunity to learn is widely considered to be the single most important predictor of student achievement" (National Research Council, 2001, p. 334).

Opportunity to learn, as applied to curriculum research, is multi-faceted. Certainly, the content included in the textbook influences the mathematics topics that students likely study (Begle, 1973). But the manner in which those topics are presented in the classroom—that is, classroom instruction—also plays a role. As Hiebert and Grouws (2007) note,

> The emphasis teachers place on different learning goals and different topics, the expectations for learning that they set, the time they allocate for particular topics, the kinds of tasks they pose ... all are part of teaching and all influence the opportunities students have to learn. (p. 379)

That is, the manner in which curriculum is enacted likely mediates the potential of a given curriculum to influence student achievement.

The purpose of this chapter is to investigate the connections between mathematics curriculum enactment and student achievement. We are guided by the following research question:

- In what ways do variations in the enactment of a mathematics curriculum influence student achievement?

In this chapter, we review representative studies that document the ways in which variations in curriculum enactment influence student achievement, but make no claim that our review is comprehensive or exhaustive. We start by considering components of curriculum enactment that many researchers consider as part of determining the extent to which a curriculum is enacted as intended by the developers, what is often called *fidelity of implementation* or *treatment/textbook integrity*. As part of this discussion, we highlight some research endeavors that have investigated differences in the ways teachers enact the curriculum to illustrate the potential for variations in student learning resulting from such differences in enactment. We then report our methodology for identifying empirical studies focused on the connection between enactment of mathematics curriculum and student learning, including documentations from several researchers on the paucity of research with this focus. After reviewing representative studies focused on the connection between curriculum enactment and student achievement, we discuss implications for future research.

COMPONENTS OF MEASURING CURRICULUM ENACTMENT

Much of the interest in research on curriculum enactment arose as a result of new curriculum materials developed, particularly in the early to mid-1990s, to instantiate the recommendations of the *Curriculum and Evaluation Standards for School Mathematics* (National Council of Teachers of Mathematics [NCTM], 1989). These materials were typically designed with philosophical and pedagogical principles quite different from those of mainline commercial publishers (Hirsch, 2007). Parents, policy makers, curriculum supervisors, and educators were interested in student achievement outcomes when using these materials. But understanding achievement in relation to these curriculum materials, or any others, required knowing whether the materials were actually used in the classroom as intended by the developers. With the requirements of the No Child Left Behind Act of 2001 that teachers use materials proven to be effective, understanding issues of implementation became even more important.

Researching issues of curriculum use or implementation was an acknowledgement that curricula are not fixed, but that enactment involves interactions among texts, teachers, and students (see Chapter 1 by Cal & Thompson for more discussion on conceptualizations of enacted curriculum and Chapter 6 by Remillard & Heck for the place of enactment within a broader conceptual framework on curriculum). For instance, Remillard (2005), in a synthesis of research about teachers' use of curriculum materials, identified a number of views about the teacher-curriculum interaction: teachers follow or subvert the text; teachers draw on the text to construct their instruction; teachers interpret the text and the authors' intentions in light of their own beliefs and experiences; and teachers collaboratively interact with the text in a dynamic relationship. Various aspects of these uses of curriculum are the basis of numerous research studies on how curriculum materials and classroom instruction interact (see Remillard, Herbel-Eisenmann, & Lloyd, 2009 for a compilation of many such studies).

In considering enactment of curriculum as part of a large-scale study of middle grades curricula, Chval, Chávez, Reys and Tarr (2009) introduced the construct of textbook integrity as "the extent to which the district-adopted textbook serves as a teacher's primary guide in determining the content, pedagogy, and nature of student activity over an identified period of time" (p. 72). They identified three components of this construct: regular use over the school year; significant use of the text to frame the content and instructional emphasis; and instructional approach consistent with the intent of the textbook developers. These components are similar to those that influenced the development of instruments to measure curriculum implementation. Such instruments consider the extent to which teachers use the textbook (e.g., what percent of the lessons are taught), adhere to the mathematical storyline (e.g., the extent to which the topics in the text are taught or the sequence is followed or the depth at which topics are taught), and teach according to the pedagogical storyline (e.g., use small groups or inquiry learning if the text is designed with that pedagogical approach in mind) (Heck, Chval, Weiss, & Ziebarth, 2012). Others have developed instruments to measure the content and use of the textbook during instruction and the style of instructional presentation (Tarr, McKnaught, & Grouws, 2012).

Using similar measures of implementation fidelity, Thompson and Senk (2010) documented implementation of textbooks by a small group of teachers who were using the same curriculum, namely one of the courses developed by the University of Chicago School Mathematics Project [UCSMP]. Among seven studies, each focusing on the implementation of a single textbook, they found that the median percent of lessons from a given textbook taught by the teachers using that text was 60%, with a range from roughly 40% to 100%. Likewise, teachers did not always assign homework problems in ways consistent with the expectations of the curriculum developers. For instance, based only on lessons taught, the percent of homework problems assigned by six teachers of UCSMP *Algebra* ranged from 48% to 100%. In one chapter of UCSMP *Algebra* in which all six teachers taught all lessons, the percent of review questions assigned varied from 0% to 95%; this variation has potential to influence student achievement significantly because UCSMP uses a modified mastery approach with continual review to develop mathematical proficiency. On a standardized achievement measure, all six algebra teachers reported having taught or reviewed the content needed for their students to answer only 16 of 32 questions (50%), again suggesting differences in their students' opportunity to learn. These results highlight the variation in implementation of the curriculum and suggest that there are likely to be differences in student achievement. However, in this report, no actual student achievement outcomes were provided.

The situation identified by Thompson and Senk occurred in the context of evaluation studies of curriculum materials prior to revision for commercial publication, what some might call efficacy studies. Similar results in terms of variability in implementation have been documented by other researchers in large-scale effectiveness studies of curricula in use in schools, specifically comparing the effectiveness of different curriculum types such as curricula developed in response to the NCTM *Standards* and typical publisher-developed textbooks (Chávez, Grouws, Tarr, Ross, & McKnaught, 2009; McNaught, Tarr, & Grouws, 2008; Tarr, Chávez, Reys, & Reys, 2006).

The variations in implementation reported by Thompson and Senk, Chávez et al. (2009), and Tarr et al. (2006) were based on use of a curriculum or textbook over an entire school year. But, as reported by Huntley and Heck (Chapter 2 of this volume) in which a single teacher implemented the same lesson in quite different ways in two class periods, variation can occur at the level of a single lesson. Based on student discourse, there appeared to be differential opportunities to learn the content of this one lesson. But once again, no specific achievement outcomes were provided.

It is the link from the implemented curriculum to the attained/achieved curriculum that is needed to understand *how* or even *if* these differences in enactment influence student achievement. We turn our attention to this link in the remainder of this chapter.

METHODOLOGY

State of Curriculum/Textbook Research

As we prepared to search for appropriate studies documenting the influence of the enacted curriculum on the achieved curriculum, we were mindful of other attempts to identify research studies explicitly linking these two levels of curriculum. For instance, in a study of the effectiveness of K-12 mathematics evaluations, a committee of the National Research Council [NRC] (2004) found only 63 studies that met minimal methodological standards containing information on outcomes and comparability of groups; only 43% of these reported measures of fidelity of implementation and only 3% used such measures to adjust for student outcomes. Similarly, O'Donnell (2008), in a review of research on measuring fidelity of implementation in relation to outcomes in K-12 curriculum research, found only five empirical studies that met all her criteria regarding information about the curriculum intervention, student outcomes, and implementation fidelity. Only two of these addressed mathematics, with one study using a computer-based curriculum intervention and one study focused on students with disabilities.

Thus, we realized we might find few studies that provided the explicit link we wanted to investigate. More recently, the lack of research studies relating curriculum and implementation to outcomes was confirmed in a survey of textbook research by Fan, Zhu, and Miao (2013) in which they identified 83 textbook studies reported in journal articles or other publications since 1990 and 28 before 1990. When categorized, they found that 34% focused on textbook analysis (i.e., the features or content), 29% on textbook comparison (i.e., similarities and differences between textbooks), and 25% on textbook use by students and teachers; the other 12% focused on either the role of textbooks in mathematics teaching and learning or other areas, such as electronic textbooks or connections between textbooks and student learning.

Search Procedures

Several criteria were used to identify studies for review and discussion. We wanted studies in which the researched curriculum was publicly available, rather than just a teacher-created unit. Because of our interest in the link between enactment and student achievement, studies needed to include one or more measures of student achievement as well as some measure of curriculum enactment. We limited our scope to works published in English involving students in K-12 schools, but included articles from peer-reviewed journals, doctoral dissertations, and technical reports.

We began the selection process by searching online education databases (e.g., Wilson Select, JStor, First Search). *Mathematics, curriculum,* and *achievement* were used in conjunction with either *enactment* or *instruction*; other searches were run using *mathematics curriculum, achievement outcomes,* and *fidelity of implementation,* with abstracts reviewed to determine whether each study clearly connected curriculum to student achievement and provided insight into the manner in which the curriculum was enacted. Enactment measures were seldom specifically described in abstracts, so further reading was required to determine a study's suitability for inclusion. We eliminated from further review any studies that did not make the curriculum explicit, provide clear measures of achievement, or give details about at least one component of enactment as previously described (i.e., extent of use, adherence to the mathematical storyline, or instructional approach).

Consistent with the limited number of studies reported by the National Research Council (2004), O'Donnell (2008), and Fan, Zhu, and Miao (2013) linking the *written curriculum* of the textbook with the *implemented curriculum* and the *achieved curriculum*, we found fewer than 20 studies that explicitly linked all three levels of the curriculum. The studies that met our criteria and are the focus of the remainder of this chapter generally fall into two broad categories: (1) studies focusing on a single curriculum and investigating differences in student achievement as a result of variations in enactment of that curriculum; and (2) comparative studies investigating the mathematics achievement of students studying from two or more different curricula.

RESULTS

We first investigate differences when the curriculum is kept constant and only the conditions of enactment vary. This enables us to provide insight into an issue raised by Kilpatrick (2003), namely the variability that exists within enactment of a single curriculum:

> Two classrooms in which the same curriculum is supposedly being "implemented" may look very different; the activities of teacher and students in each room may be quite dissimilar, with different learning opportunities available, ... and different outcomes achieved. (p. 473)

We then consider research studies in which two or more curricula are compared, but limit our review to studies that provide information about enactment and achievement results.

Connections Between Enactment and Achievement From Studies of a Single Curriculum

In this section, we review several noncomparative studies focusing on different aspects of implementation. We have grouped these studies according to levels of curriculum coverage, coherence of instructional approach with the curriculum's pedagogical stance, and strength of implementation.

Studies Focused on Varying Levels of Curriculum Coverage

It is natural to expect that the amount of a curriculum that is taught or that students have an opportunity to study should influence their achievement, at least if the achievement is related to the curriculum. The study reported here uses a review of student workbooks, rather than teacher reports of lessons taught, to gauge curriculum coverage.

Cueto, Ramirez, and Leon (2006) evaluated student workbooks at the end of the school year for sixth-grade Peruvian students from one class at each of 22 randomly selected public elementary schools. Students had spent the year studying the national curriculum. Researchers measured curriculum coverage using the proportion of exercises completed by the best two students in each class within each content strand, the cognitive demand of each exercise, and the level of teacher feedback on students' completed work. These components were combined to form an overall opportunity-to-learn factor.

Mathematics achievement was assessed via the national achievement test. The results of hierarchical linear modeling found that opportunity to learn was a significant predictor of student achievement, as might be expected. However, when the three components of opportunity to learn were analyzed separately, only content coverage was significant; neither the cognitive demand of the tasks nor teacher feedback were significant predictors of achievement. More detailed study of content coverage found that only 44% of the tasks in the workbooks were solved by the end of the year and 80% of the teachers taught topics not part of the national curriculum.

This study raises a number of questions. Why did teachers teach so many topics not in the curriculum? How many units from the national curriculum were taught? How did the percent of tasks completed and the content of those tasks relate to the national exam? Despite the relatively low percent of completed workbook tasks and the high percentage of topics taught that were not in the national curriculum, these issues of implementation may be immaterial as long as the content on the national exam was taught.

Studies Focused on the Coherence of Instructional Practice With Curricular Philosophy

Curriculum development begins with the authors' theoretical framework—their beliefs about what mathematical ideas and experiences are important for students to learn, their perspectives on how children learn mathematics, and beliefs about how teachers should teach mathematics.

The theoretical framework is inextricably woven throughout the written curriculum in how it is designed and how it is intended to be enacted; sometimes the framework is explicitly communicated to users of the curriculum (e.g., Hirsch, 2007) and sometimes it is not. Research on the coherence of instructional practice with curricular philosophy provides a means to evaluate how differences in implementation impact the ability of a curriculum to yield the outcomes for which it was designed. If the philosophical stance is not available in some resource, researchers may need to infer the philosophy from their own reading of the curriculum or interview the curriculum authors to develop a means of assessing instructional coherence (Tarr, McKnaught, & Grouws, 2012).

Schoen, Cebulla, Finn, and Fi (2003) attempted to relate teacher behaviors aligned with the curriculum's design and various demographic variables to student achievement. They worked with 40 teachers who were field-testing Course 1 of the *Core-Plus Mathematics Project* (CPMP) curriculum during their students' first exposure to the curriculum. In this curriculum, mathematics concepts are developed in context using interwoven content strands. Student exploration, sense-making, and flexible thinking are promoted, with students expected to work collaboratively in various formats.

Evidence of curriculum enactment was gathered using classroom observations and both mid-year and end-of-year teacher surveys. Classroom observations noted practices aligned to the curriculum, such as class organization (i.e., whole-class, small group), open-ended questioning, students' monitoring their own work, students' collaborative behaviors, and use of manipulatives and technology tools. The mid-year survey focused on teachers' perceptions of their classroom practices, including the portion of class time spent with students working in different formats, use of curriculum features, assessment practices, and the extent to which they supplemented or revised the curriculum materials. The end-of-year survey measured teachers' concerns about, and preparation for, curriculum implementation.

Students ($n = 1466$) were pre- and post-tested using parallel forms of the *Iowa Test of Educational Development*, a test that measures conceptual understanding, problem solving, applications, and quantitative thinking. The percent of students in each school receiving free/reduced-price lunch and the percent of under-represented minorities (African American, Native American, and Hispanic) were highly negatively correlated with students' mean pretest scores ($r = -0.63$ and $r = -0.70$, respectively). However, the adjusted mean posttest scores were not significantly correlated with either of these variables; none of the demographic variables explained variance in achievement when pretest scores were controlled.

The strongest predictor of student achievement among teacher variables was whether or not the teacher completed the two-week summer workshop about the curriculum. The teacher observation results indicate that adjusted mean student achievement was higher for teachers whose teaching practices were aligned with the CPMP developers' recommendations. Data from the mid-year teacher survey indicated that teachers' high expectations on homework, high grading standards, and the degree to which the curriculum was implemented as designed, were significantly associated with student achievement. The students of teachers who reported the lowest levels of supplementing, replacing, or revising the curriculum and assessments experienced the highest achievement gains.

Similarly, Jong and colleagues (2010) observed the relationship between reformed teaching practices and student achievement after a single 4-6 week unit of curriculum enacted by beginning elementary school teachers. Researchers used the Reformed Teaching Observation Protocol (RTOP), an instrument that measures active learning, inquiry-based instruction, and problem-solving strategies, to determine the extent to which teachers' enactment was consistent with the goals of the two study curricula, *Investigations* (K-5), and *Connected Mathematics Project* (grade 6). The participants were 22 teachers in a large, urban school district with a high percentage of students of color, English language learners, and students receiving free/reduced lunch.

Each teacher was observed twice during the unit by trained observers who used the RTOP to evaluate lesson design and implementation, the type of knowledge that was the lesson focus, and the classroom culture. Student achievement on the unit was measured via a district-developed test containing seven multiple-choice and three constructed-response items. The correlation between teachers' RTOP scores and their pupils' posttest content scores was 0.56 ($p < 0.05$), indicating that teaching practices aligned with the respective curricula were positively and significantly related to students' mathematics learning.

Studies Focused on Strong Versus Weak Implementation of a Curriculum

Implementation of a curriculum is not typically an all or nothing affair. Teachers may strongly embrace the curriculum, implementing it as intended with any adaptations consistent with the philosophical and pedagogical stance of the curriculum developers. Others may implement the curriculum in terms of using the materials as a basis for instruction, but make adaptations that are in conflict with the curriculum's design, or supplement the curriculum in ways that potentially undermine the curriculum's goals. Still others may think they are implementing the curriculum, but misunderstand its philosophy of instruction or feel that they can maintain a conflicting but familiar philosophy of instruction.

Briars and Resnick (2000) researched the relationship between the strength of fourth-grade teachers' enactment of the *Everyday Mathematics* (*EM*) curriculum and student achievement. Students were taught by 111 teachers in 38 schools, all who used EM and participated in the Pittsburg Reform in Mathematics Education Project (PRIME), a district-wide effort to improve student achievement. The project offered in-class support via demonstration lessons, joint planning among teachers, and coaching, together with summer and after-school professional development.

Teachers were rated by their school's PRIME demonstration teacher according to the nature of their curriculum implementation. *Strong Implementers* used all of the curriculum components and provided student-centered instruction. *Weak Implementers* were either not using *EM* or were using it so little that their instruction was hardly distinguishable from traditional instruction. Teachers who fell between these two levels were excluded from the study. Schools with all teachers in third and fourth grade classified as Strong Implementers were called Strong schools ($n = 8$); those with all but one or two teachers who were Weak Implementers were classified as Weak schools ($n = 3$). Strong and Weak schools were matched according to measures of race and socio-economic status, resulting in three matched pairs. Unmatched Strong schools (Other Strong) were included in some phases of the analysis.

Researchers used prior achievement data from the *Iowa Test of Basic Skills* [ITBS] to ensure that students in paired schools were not significantly different in achievement at the beginning of the study. Two outcome measures were used: the *New Standards Mathematics Reference Examination* [NSMRE] which assesses skills, concepts, and problem solving on three subtests aligned with the *EM* curriculum; and the survey battery of the ITBS, a norm-referenced test, to ensure that implementation of this curriculum would not negatively impact student's achievement on more traditional skills. Data were collected for three years, but it was only in the third year of the study that fourth-grade students had experienced the *EM* curriculum beginning in kindergarten.

Over the final two years of the study, the percent of students who achieved the standard on the NSMRE Skills subtest rose by over 20% and on the Concepts and Problem Solving subtests by an average of 10%; there was a sharp drop on all three subtests in those who scored below the standard. In addition, student gains on the NSMRE did not come at the expense of more traditional skills as measured by the ITBS. More students scored at or above the 50th and 75th percentiles in the final year of the study, and fewer students scored below the 25th percentile.

To consider the influence of implementation, in the third year comparisons were made of achievement scores between students in Strong and Weak schools. On the NSMRE, more than twice as many students in Strong schools as in Weak schools achieved the standard for Skills, more than five times as many achieved the standard for Concepts, and more than four times as many achieved the standard for Problem Solving. For those below the standard, half as many students in Strong schools compared to Weak schools were below the norm in problem solving, less than one-third as many were below in Concepts, and one-fourth as many were below in Skills. A similar pattern of results was found when comparing the ITBS scores of students in Strong versus Weak schools; more than twice as many students in Strong schools compared to Weak schools were at or above the 75th percentile and less than half as many were below the 25th percentile.

To ensure that achievement differences found were due to the curriculum and the fidelity of implementation rather than general teacher quality, the researchers compared the achievement scores of students in Strong and Weak schools across all three years. In the third year, there was a spike in NSRME Concepts scores for students in Strong schools; while the scores of students in Strong schools improved every year of implementation, the scores of students in Weak schools remained constant. For Skills, the same pattern emerged except students in Weak schools improved in the third year, but not nearly as much as those in Strong schools. For Problem Solving, the pattern was similar except that there was not a significant increase in scores between years one and two for students in Strong schools. For all three subtests, students in Strong schools performed better than those in Weak schools in year one, meaning that their teachers may have been more skilled, but the real differences in achievement came after years of Strong implementation.

The results comparing the performances of African American and white students in Weak schools versus Strong and Other Strong schools provide evidence that when a curriculum is taught in accordance with its design, the achievement gap can be narrowed. On the NSMRE Skills subtest, African American students in both Strong and Other Strong schools performed much better than their counterparts in Weak schools. In the Strong schools, there was no significant difference in performance between African American and White students. For the Concepts and Problem Solving subtests, African American and white students in Strong and Other Strong schools scored significantly better than their counterparts in Weak schools.

A similar pattern of results in relation to the level of implementation was found in a study by Pierce and colleagues (2011). They documented the influence of varying levels of curricular implementation on clustered groupings of gifted third-grade students in urban elementary schools. In this district, all gifted students in each grade were clustered in a single classroom with their non-gifted peers, with a cluster defined as 3-10 gifted students in a classroom of 20-25. Although the study was multi-year, only the Year 1 report compares student achievement ($n = 161$) based on varying levels of enactment.

Teachers were categorized as *implementers* or *non-implementers* based on their attendance at a summer training institute, observations of their teaching, and self-reports of their implementation of the instructional materials and procedures. For two units of 9-weeks each, the teachers replaced their regular curriculum with the treatment curriculum, using *Into the Unknown* (Bippert & Vandling, 2001) for algebra and *Math By All Means* (Rectanus, 1994) for geometry. Student achievement and gains were measured via unit pretests and posttests.

For the algebra unit, a repeated measures ANOVA showed a statistically significant main effect for implementation status and gifted status. As might be expected, the gifted students outperformed their non-gifted peers, and students in classrooms with teachers who fully implemented the curriculum outperformed those with teachers who did not. Of particular interest was the interaction effect by implementation status. Student growth in achievement was influenced more strongly by level of implementation of the curriculum than by gifted status. Although gifted students outscored non-gifted students on the pretest, gifted students in classes whose teachers did not implement the curriculum scored lower on the posttest than the non-gifted students in classes in which the curriculum was implemented. Similar results were found for the geometry unit with curriculum implementation having a significant effect on students' achievement growth. All students in the classrooms of implementing teachers experienced learning gains for both units, whereas achievement results for students in the classrooms of non-implementing teachers were mixed.

Connections Between Enactment and Achievement in Comparative Studies

Several research studies have been conducted within the context of efficacy or effectiveness studies related to curricula developed as part of curriculum projects, often in response to the NCTM *Standards*. These studies were often conducted to satisfy critics concerned that studying from curricula with an alternative format (e.g., modules or integrated content rather than subject-specific content) or alternative pedagogical expectations (e.g., extensive work within small groups or advanced use of technology) would lead to reduced mathematical proficiency. Different teams of researchers have investigated achievement when students study from these curricula in comparison to typical commercial curricula, with various components of enactment included in their research designs. We report results from several research teams.

Middle School Mathematics Study

Tarr and colleagues (2008) studied the achievement of a large group of middle grades students over 3 years in 10 different schools who were studying from either a publisher-developed textbook series or one of three textbooks developed with funding from the National Science Foundation (NSF) (*Connected Mathematics*, *MathThematics*, or *Mathematics in Context*). Textbooks had been adopted by the school district, with all schools having implemented the curriculum for at least one year prior to participation in the study.

The research team collected a range of implementation data from teachers to determine how they used their respective curriculum materials to plan and enact their instruction: teacher questionnaires, textbook-use diaries in which teachers recorded specific information about the use of the textbook for three 10-day intervals, and a table of contents record documenting the lessons taught. In addition, classroom observations were conducted to determine the extent to which the classroom environment modeled practices recommended by the NCTM *Standards* and aligned with the philosophical approach of the NSF-funded curricula. Observers also documented how teachers and students used the textbook during instruction, how it influenced the lesson content and presentation, and what homework was assigned. Thus, this study collected data to address implementation of the pedagogical storyline in addition to the extent of use and adherence to the mathematical storyline. These different aspects of implementation were combined to develop an implementation index.

Students' mathematics achievement was measured via two tests: the *TerraNova Survey*, a multiple-choice test that assesses knowledge across all five content domains identified in the NCTM Standards; and the *Balanced Assessment in Mathematics*, a criterion-referenced constructed-response test to assess reasoning, problem solving, and communication. The research team was particularly interested in the extent to which teachers' implementation in terms of curriculum type or classroom learning environment would influence student achievement.

The researchers determined that teachers using both types of curriculum were strong implementers of their curriculum and utilized their textbooks appropriately. Using hierarchical linear modeling, no differences in student achievement were noted for either of the two tests in relation to teachers' implementation index. However, curriculum type and level of standards-based learning environment did interact in terms of student achievement on the *Balanced Assessment*. When students studied from one of the NSF-funded curricula and were in classrooms in which teachers developed a moderate or high standards-based learning environment, achievement was positively influenced; no significant relationship was found for NSF-funded curricula and low levels of a standards-based learning environment or for the publisher-developed textbooks and the learning environment. On the multiple-choice *TerraNova*, the interaction of learning environment and curriculum type was not significant.

The researchers noted the complexities of determining implementation, and the variability in implementation that often occurred among teachers in the same school. They hypothesized that the interaction observed between curriculum and the learning environment only on the *Balanced Assessment* was perhaps due to the coherence between this assessment, a standards-based learning environment, and the philosophical stance of the NSF-funded curricula. That is, when instruction occurred in a manner consistent with that held by the curriculum developers and this philosophical stance was reflected on an assessment, students achieved at a somewhat higher level.

Study From the Middle School Talent Development Program

Balfanz, MacIver, and Byrnes (2006) engaged in a study of an attempt to enhance middle grades mathematics achievement in one low-income district. They worked under the assumption that improving achievement at the high school level starts in the middle grades.

At the time of the study, various teacher variables resulted in a highly repetitive course of study across grades; most middle grades teachers in the district were elementary certified with no coursework in middle grades mathematics, did not choose to teach mathematics, and selected their own curriculum. In the schools serving as comparisons, only one-fourth of the teachers taught the same subject at the same grade level for two consecutive years. In addition to instructional issues, students entered the middle grades significantly below grade level in mathematics.

During the four years of the study period, the school district pursued an ambitious reform agenda, and all schools had two-year growth targets. The faculty of three schools voted to enact the Talent Development (TD) Middle School Mathematics Program as part of a whole-school reform effort in order to meet district goals. The program targets high poverty middle schools and integrates reforms in school organization, curriculum, professional development, school climate, teacher-student interaction, and student support. For the study schools, *EM* was adopted for grades 5 and 6, UCSMP *Transition Mathematics* was used for grade 7, and UCSMP *Algebra* was used for grade 8. Although schools phased in the curriculum differently, all schools offered all students the same curriculum by the third year of the study.

The three TD schools were matched with three comparison schools in the district based on poverty status, racial composition, and past performance. Across all six schools ($n = 1,174$ students, 36 teachers), the percent of students below grade level ranged from 71-86%. Each school had only one teacher certified in secondary mathematics with the remaining teachers certified at the elementary level. Teachers in the TD schools received strong curriculum-specific professional development to adapt to the new curriculum, with 3 days of training in the summer followed by monthly 3-hour workshops; facilitators of the professional development modeled all elements of the curriculum as well as the mechanics of having students work with manipulatives. Similar training was provided in the second year of the study; in the third year, two to three teachers per school were trained to be teacher leaders as on-site trainers to provide implementation support at the school level. Nearly 80% of teachers participated in the professional development, and about two-thirds completed the full 36 hours per year for the first two years.

The level of curriculum enactment was measured in two ways; students were surveyed about the frequency with which they experienced recommended instructional practices, and the curriculum coaches rated teachers' frequency and intensity of those same recommended practices. The overall implementation index included teacher experience teaching the adopted program, use of pedagogy, curriculum coverage, and professional development. One half of classrooms fell in the medium-high range of implementation, one fourth in the high range, and one fourth in the medium-low range. The content coverage goal was 6-8 units per year; however, because of various issues, each TD school covered between 3.8 and 4.7 units per year. Of the nine identified and recommended reform teaching practices, teachers in the TD schools used an average of 5.2 recommended practices frequently versus 4.7 for teachers in the control schools, a difference that was statistically significant.

Student achievement was measured via the Stanford 9 problem solving test two to three times during the four years of the study. None of the schools had statistically significant differences in achievement scores prior to the study. The state's high-stakes test was also administered during the spring of fifth and eighth grade, so researchers looked at *value added* by the curriculum for the last three years of TD (grades 6-8).

Students in TD schools scored significantly higher on the Stanford 9 and the state's high-stakes test than students in the comparison schools. The ratio of students above the 50th percentile rose from 6% to 11% for TD schools and from 5% to 7% for control schools. Thirty-two percent of TD students across all prior achievement levels gained ten or more percentile points versus 22% for comparison students. These differences were significant for all low-achieving students, but were not significant for those slightly below average or higher.

For the purpose of this chapter, the most interesting results came from analyzing the relationship between achievement and a blend of level, context, and implementation for all classrooms within the TD schools. Implementation for each classroom was measured via a comprehensive index that included the following variables: amount of TD curriculum covered; years teaching TD; use of TD pedagogy; and hours of professional development attended. That score was then reduced based on the number of roadblocks to implementation that were encountered and the level of teacher turnover in the class. Within the TD schools, students in classrooms with higher levels of program implementation averaged significantly higher achievement gains.

Achievement gains were realized in spite of high teacher turnover, inadequate teacher preparation for middle grade mathematics, high poverty, and students' low prior mathematics achievement. The implementation of comprehensive mathematics reforms resulted in significant achievement gains across multiple classrooms in multiple schools across multiple years and all achievement levels, and higher levels of implementation netted higher student gains.

Study From Curriculum Implementation as Part of the Urban Systemic Initiative

McCaffrey and colleagues (2001) examined the degree to which teachers' use of instructional practices that aligned with their curriculum influenced student achievement. The sample consisted of nearly 5,000 10th-grade students in 226 classes from 182 teachers in 27 schools in one large, urban, high-poverty district. At the time of the study, the district was participating in the National Science Foundation's Urban Systemic Initiative with district-wide reform focused on inquiry-based instruction, reasoning and problem solving, and connections among mathematics and science topics. Students could complete an integrated, inquiry-based curriculum (either the *Interactive Mathematics Program* (IMP) or *College Preparatory Mathematics* (CPM)) or a traditional commercial publisher sequence of algebra I, geometry, and algebra II/trigonometry. Both of the integrated curricular programs are problem-based, focus on conceptual understanding, and use cooperative learning extensively, even though the CPM is offered within a traditional sequence of courses.

Researchers collected student achievement and demographic data, including race/ethnicity, gender, language history, socioeconomic status, and history of repeating a grade. Comparability of groups was based on prior achievement as determined by the Stanford-9 open-ended test completed when students were in ninth grade. Both the multiple-choice and open-ended sections of the Stanford 9 were used in tenth grade as outcome measures. These tests, like the integrated curricula, are designed to align with the NCTM *Standards* and heavily emphasize problem solving.

In addition, researchers collected data on teacher practices via a questionnaire with two scales; the *reformed practices* scale queried teachers about how often they used student-led discussions, investigations, problem solving, and portfolio-based assessments; the *traditional practices* scale asked about the use of more conventional approaches, such as textbooks, worksheets, and multiple-choice tests. Survey data were validated by comparing teacher responses to classroom observations, teacher interviews, and a review of artifacts, with moderately high correlation evident.

Teachers who taught integrated courses used reform practices significantly more often and used traditional practices significantly less often than teachers who taught the traditional course sequence. The correlation between the two scales was low. For teachers who taught the integrated curricula it was -0.32, i.e., those who reported more of one scale tended to report less of the other. For those who taught the traditional sequence, the correlation was 0.27, i.e., those who reported using one kind of practice more also tended to report using the other type of practice more. Teachers who taught both traditional and integrated courses tended to have a higher reform practice score for their integrated classes than for their traditional classes; they apparently changed their teaching practices to some extent to be aligned with the philosophy of the curriculum.

Teachers' more frequent use of reform practices was associated with higher test scores on both portions of the Stanford 9 for students in integrated courses. For students in traditional courses, neither the reform practice scale nor the traditional practice scale was a significant predictor of achievement. There was no relationship between the use of traditional practices and test scores for any courses. The demographic variables suggested important implications for the achievement gap; poverty was a significant predictor of achievement in the traditional classes, but was not significant in the integrated classes. For both course types, females scored lower than males, but the difference was only significant in the traditional courses.

Student scores on both the traditional multiple-choice and open-ended portions of the Stanford 9 were higher in integrated courses taught via reform practices. Overall, the researchers suggest that reform practices may be more effective when used in courses designed to be taught with instruction aligned to the *Standards*.

Studies From the University of Chicago School Mathematics Project [UCSMP]

Thompson and Senk (2001) report on achievement of students studying from UCSMP *Advanced Algebra*, a course expected to follow a geometry course in sequence and be equivalent to a second year of algebra study. The study used matched pairs of classes at the same school, with students in one class using UCSMP *Advanced Algebra* and students in the other class using the commercially-published second-year algebra textbook already in place. Additionally, students in the matched pairs were themselves matched on the basis of a pretest of prerequisite background knowledge. Implementation measures included classroom observations, teacher interviews, end-of-year teacher questionnaires, and teachers' reports of the extent to which content on the posttests was taught or reviewed during the year.

On some aspects of instructional approach there were reported differences among teachers. For instance, UCSMP teachers reported spending less time lecturing or providing demonstrations and more time reviewing homework. However, on other measures there was little difference, with teachers in both curriculum groups expecting students to write solutions and regularly complete homework. Availability and frequency of use of graphing calculators were similar between classes at some schools and quite different at others.

Achievement on the posttest was reported in light of the opportunity to learn the content of the items on the posttest—a variation of enactment measures considering extent of use and adherence to the mathematical storyline. Among the eight-matched pairs of classes in four schools, the eight teachers reported students had an opportunity to learn the content for from 47 to 100% of the items. Achievement differences among class means were significant for four pairs, and a matched-pairs t-test indicated an overall significant difference between students studying from the UCSMP and non-UCSMP curricula.

Thompson and Senk investigated the achievement results in two other ways that controlled for one aspect of implementation by controlling for opportunity to learn the posttest items. For each school, a "fair test" was compiled using only those posttest items for which both teachers at the school reported students had an opportunity to learn the needed content. The number of items on these fair tests ranged from 17 out of 36 items at one school, to 25 items at a second school, to 26 items at the other two schools. Controlling for opportunity to learn at the school level yielded differences in class means for three pairs. Again, there was an overall significant difference in achievement between students studying from the two curricula.

However, when opportunity to learn was controlled across all the classes in the study, there were only 15 of the 36 items for which all teachers reported having taught or reviewed the necessary content, and most were items focusing on procedural skills. Once implementation was controlled at this level, any significant differences in achievement between students studying from the different curricula disappeared. Limiting the analysis of achievement in this way ignores potential curricular differences (e.g., the mathematical storylines) between textbooks, though instructional differences are not ignored. Perhaps the absence of significant differences should not be unexpected when implementation of assessed content is controlled in this manner.

In a follow-up study of the third edition of this textbook, Senk, Thompson, and Wernet (2014) investigated the influence of curriculum and teachers' implementation on students' achievement with functions, a major topic of study for students at this level and a focus of the curriculum recommendations in the *Common Core State Standards for Mathematics* (Council of Chief State School Officers, 2010). Matched-pairs of classes in five schools participated in this study, with 20 classes taught by 10 teachers. The percent of textbook lessons related to functions that were reported as taught ranged from 31 to 86%; in all but one school, the percent of function lessons taught was roughly comparable by teachers using two different curricula.

However, when reporting on the opportunity to learn the content on the posttests—one measure of implementation—major differences were evident. Only 5 of 37 function items, a mere 13.5%, were reported as taught or reviewed by all teachers, all with a focus on interpreting functions. Achievement results were reported in relation to the opportunities to learn. Regression analyses, using implementation factors of function lesson coverage and posttest opportunity to learn, found that students studying from the UCSMP curriculum had a 5-6% advantage on two of the three posttests, with no significant difference in achievement of students studying from different curricula on the other posttest. Posttest opportunity to learn was a significant predictor of achievement on all three posttests, with function lesson coverage a significant predictor of achievement on two of the posttests.

Overall, these results from the UCSMP studies indicate the importance of providing the implementation context of curriculum exposure and coverage when interpreting achievement results. Apparent differences in achievement often disappeared when opportunity to learn was held constant, although again such controls ignore curricular differences between textbooks.

Comparing Options in Secondary Mathematics: Investigating Curricula [COSMIC]

At the high school level, students often study mathematics in subject-specific courses, such as algebra I, geometry, and algebra II. But with the advent of the NCTM *Standards* (1989), some curriculum developers generated curricula with an integrated sequence and a philosophical stance in which students would work extensively in small-cooperative groups to explore mathematics. Grouws et al. (2013) and Tarr et al. (2013) report results of student achievement when studying from one of these two curricular pathways. As part of their large scale research study, they developed a series of instruments to collect data on classroom implementation of each type of curriculum in order to use statistical modeling to relate implementation measures to curriculum type and student achievement.

In this longitudinal study, students either used subject-specific curriculum materials developed by commercial publishers or used the integrated NSF-funded *Core Plus* curriculum. Both curricular pathways were available in the same school, with students self-selecting into the curricular pathway of choice. Comparability of groups was determined from state-mandated eighth-grade test results.

Student achievement data were determined from several measures, with the actual content of items appropriate for the given year of study. The COSMIC group developed two tests appropriate for each year of the study: a test of common objectives, and a problem-solving and reasoning test. In addition, the standardized *Iowa Test of Educational Development* was used to assess computational and problem-solving skills in a range of contexts on a nationally recognized assessment.

Teachers' implementation of their respective curricula was documented using many of the same types of measures as in the Middle School Mathematics Study. An initial teacher survey collected demographic information, teacher beliefs, and participation in professional development, and a mid-course teacher survey collected information about instructional practices, use of technology, and assessment practices. A table of contents record documented whether a lesson from the textbook was taught, and if so, with what level of supplementation, and classroom observation protocols documented the fidelity of content and presentation implementation. These data were used to create a series of curriculum implementation factors.

In analyzing data from the first year, Grouws et al. (2013) found that teachers from both curriculum types taught about the same percentage of lessons from their textbook (78%) and generally taught without additional supplementary resources. The researchers report that teachers tended to implement the content of their respective textbooks but presented the content with more adaptability than might have been intended by the curriculum developers. There were some differences in instructional practice, with teachers using the integrated curriculum encouraging reasoning, using small cooperative groups, and allocating more time for lesson development than teachers using the subject-specific curricula.

When analyzing first-year student achievement using hierarchical linear modeling, students using the integrated curriculum scored significantly higher than those using the subject-specific curriculum on all three posttests. Greater teacher experience and more opportunity to learn were also significantly related to achievement on the test of common objectives and the standardized test, but not on the problem solving and reasoning test. However, implementation as measured by the classroom learning environment was not significantly related to any of the three achievement measures.

On the assessments for the second year of the study, Tarr et al. (2013) reported that opportunity to learn as determined by the percent of the textbook used was a significant predictor of achievement on the three assessments. Although students from both types of curricula performed comparably on the two tests developed by the project, students in the integrated curricula performed better on the standardized measure of achievement. In terms of implementation, the fidelity of implementation factor was not significant in predicting achievement, but the classroom learning environment was able to predict achievement in some of the developed models.

Overall, the COSMIC researchers provided insight into achievement with different curriculum types when data document that the curricula were generally implemented as intended. They do, however, point to the complex nature of studying implementation, and the many interactions that may hinder finding significant relationships.

DISCUSSION

Our goal in the review presented in this chapter was to investigate the influence of the enacted curriculum on student achievement. It seems evident that whether student achievement should be attributed to the effects of a curriculum, either positively or negatively, is dependent upon the extent to which the curriculum was used. Furthermore, it is not just usage of the curriculum that is important, but usage that is aligned with the pedagogical stance of the curriculum's authors. As noted by the National Research Council (2004),

Evaluation should present evidence that provides reliable and valid indicators of the extent, quality, and type of the implementation of the materials. At a minimum, there should be documentation of the extent of coverage of curricular material.... Evaluators are advised to provide reports on other implementation factors ... reports of the assignment of students and differential impacts, instructional quality and type, the beliefs and understandings of teachers and students, ... time and resource allocations. (p. 194)

Much of the work related to identifying and measuring curriculum implementation has been done within the context of research about the effectiveness of the National Science Foundation-funded curriculum materials or similar curriculum materials (e.g., UCSMP). Perhaps this was to be expected because initial research needed to document that student achievement when using these materials would be at least as good, if not better, than when students studied from the curriculum previously used that was familiar to parents, educators, and policy makers.

The results presented here, as well as those presented elsewhere (e.g., Senk & Thompson, 2003), have consistently demonstrated that students do not appear to be harmed when using these curriculum materials. It seems time to focus much more on answering questions along the lines of "How does curriculum X work with a particular group of students taught by teachers with given characteristics?" or "How does curriculum X work with a particular group of students who have a given level or deficit of prerequisite knowledge?" or "What supports (e.g., mathematical knowledge, classroom management skills, instructional approaches) are needed in order for curriculum X to be implemented by a given group of teachers in ways that maximize the effectiveness of the curriculum on student achievement?" Such questions will likely require a shift in focus from studies comparing the effectiveness of curriculum X to curriculum Y to studies comparing the effectiveness of curriculum X when implemented by teachers W, Y, and Z with students having characteristics A, B, and C.

The research reviewed in this chapter indicates that much work has already been done in developing instruments to assess curriculum implementation (see Chapter 5 by Ziebarth, Fonger, and Kratky). Some measures are easy to collect and document at scale, such as Table of Contents logs, teacher diaries, or teacher questionnaires about instructional practices, particularly with technology that could enable many such reports to be completed electronically with data immediately transferred to a database. But assessing the classroom learning environment in an objective manner, as done by Grouws et al. (2013) and Tarr et al. (2013), poses challenges for research at scale because of the human and financial resources required for large numbers of observations, yet their research highlights the importance of such measures.

Although opportunity to learn, in terms of content coverage and the use of the textbook to focus lessons, is a significant predictor of achievement, other implementation indices or factors are sometimes significant and sometimes not, or significant only with parsimonious models using HLM. As the COSMIC researchers note, many of the implementation factors are intertwined with each other, perhaps making it difficult to obtain significance when they compete for variance in the achievement results. More research is clearly needed to tease apart these implementation factors. With more studies to investigate variations in enactment within a single curriculum resource (i.e., textbook), perhaps more detailed study of specific features of a curriculum can be analyzed to gain a better understanding of the relationship between implementation and achievement.

REFERENCES

Balfanz, R., MacIver, D. J., & Byrnes, V. (2006). The implementation and impact of evidence-based mathematics reforms in high-poverty middle schools: A multi-year study. *Journal for Research in Mathematics Education, 37*(1), 33-64.

Ball, D. L., & Cohen, D. K. (1996). Reform by the book: What is—or might be—the role of curriculum materials in teacher learning and instructional reform? *Educational Researcher, 25*(9), 6-8.

Begle, E. G. (1973). Some lessons learned by SMSG. *Mathematics Teacher, 66,* 207-214.

Bippert, J., & Vandling, L. (2001). *Into the unknown.* Fort Atkinson, WI: Interaction Publishers.

Briars, D. J., & Resnick, L. B. (2000). *Standards, assessments – and what else? The essential elements of standards-based school improvement.* (Technical Report No. 528). Retrieved from Center for the Study of Evaluation http://www.cse.ucla.edu/products/Reports/TECH528.pdf

Cai, J., Wang, N., Moyer, J. C., Wang, C., & Nie, B. (2011). Longitudinal investigation of the curricular effect: An analysis of student learning outcomes from the LieCal Project in the United States. *International Journal of Educational Research, 50*(2), 117-136.

Cal, G., & Thompson, D. R. (2014). The enacted curriculum as a focus of research. In D. R. Thompson & Z. Usiskin (Eds.), *Enacted mathematics curriculum: A conceptual framework and research needs* (pp. 1-19). Charlotte, NC: Information Age Publishing.

Chávez, Ó., Grouws, D. A., Tarr, J. E., Ross, D. J., & McKnaught, M. D. (2009, April). *Mathematics curriculum implementation and linear functions in secondary mathematics. Results from the comparing options in secondary mathematics project.* Paper presented at the Annual Meeting of the American Educational Research Association, San Diego, CA.

Chval, K. B, Chávez, Ó., Reys, B. J., & Tarr, J. (2009). Consideration and limitations related to conceptualizing and measuring textbook integrity. In J. T.

Remillard, B. A. Herbel-Eisenmann, & G. M. Lloyd (Eds.), *Mathematics teachers at work: Connecting curriculum materials and classroom instruction* (pp. 70-84). New York: Routledge.

Council of Chief State School Officers. (2010). *Common core state standards for mathematics.* Washington, DC: Author. Retrieved from http://www.corestandards.org/

Cueto, S., Ramirez, C., & Leon, J. (2006). Opportunities to learn and achievement in mathematics in a sample of sixth grade students in Lima, Peru. *Educational Studies in Mathematics, 62*, 25-55.

Fan, L., Zhu, Y., & Miao, Z. (2013). Textbook research in mathematics education: Development status and directions. *ZDM: The International Journal of Mathematics Education, 45*(5), 633-646.

Grouws, D. A., & Smith, M. (2000). NAEP findings on the preparation and practices of mathematics teachers. In E. A. Silver & P. A. Kenney (Eds.), *Results from the seventh mathematics assessment of the National Assessment of Educational Progress* (pp. 107-139). Reston, VA: National Council of Teachers of Mathematics.

Grouws, D. A., Tarr, J. E., Chávez, Ó., Sears, R., Soria, V., & Taylan, R. D. (2013). Curriculum and implementation effects on high school students' mathematics learning from curricula representing subject-specific and integrated content organizations. *Journal for Research in Mathematics Education, 44*(2), 416-463.

Heck, D. J., Chval, K. B., Weiss, I. R., & Ziebarth, S. W. (2012). Developing measures of fidelity of implementation for mathematics curriculum materials enactment. In D. J. Heck, K. B. Chval, I. R. Weiss, & S. W. Ziebarth (Eds.), *Approaches to studying the enacted mathematics curriculum* (pp. 67-87). Charlotte, NC: Information Age Publishing.

Hiebert, J., & Grouws, D. A. (2007). The effects of classroom mathematics teaching on students' learning. In F. K. Lester (Ed.), *Second handbook of research on mathematics teaching and learning* (pp. 371-404). Charlotte, NC: Information Age Publishing.

Hirsch, C. R. (Ed.). (2007). *Perspectives on the design and development of school mathematics curricula.* Reston, VA: National Council of Teachers of Mathematics.

Huntley, M. A., & Heck, D. J. (2014). Examining variations in enactment of a grade 7 mathematics lesson by a single teacher: Implications for future research on mathematics curriculum enactment. In D. R. Thompson & Z. Usiskin (Eds.), *Enacted mathematics curriculum: A conceptual framework and research needs* (pp. 21-45). Charlotte, NC: Information Age Publishing.

Husen, T. (Ed.). (1967). *International study of achievement in mathematics: A comparison of twelve systems. Volumes I and II*. Stockholm, Sweden: Almqvist & Wiksell.

Jong, C., Pedulla, J. J., Reagan, E. M., Salomon-Fernandez, Y., & Cochran-Smith, M. (2010). Exploring the link between reformed teaching practices and pupil learning in elementary school mathematics. *School Science and Mathematics, 110*(6), 309-326.

Kilpatrick, J. (2003). What works? In S. L. Senk & D. R. Thompson (Eds.), *Standards-based school mathematics curricula: What are they? What do students learn?* (pp. 471-488). Mahwah, NJ: Lawrence Erlbaum.

Love, E., & Pimm, D. (1996). "This is so": a text on texts. In A. J. Bishop, K. Clements, C. Keitel, J. Kilpatrick, C. Laborde (Eds.), *International handbook of mathematics education* (pp. 371-409). Dordrecht, Netherlands: Kluwer Academic Publishers.

McCaffrey, D. F., Hamilton, L. S., Stecher, B. M., Klein, S. P., Bugliari, D., & Robyn, A. (2001). Interactions among instructional practices, curriculum, and student achievement. The case of standards-based high school mathematics. *Journal for Research in Mathematics Education, 32*(5), 493-517.

McKnaught, M. D., Tarr, J. E., & Grouws, D. A. (2008, March). *Assessing curriculum implementation: Insights from the comparing options in secondary mathematics investigating curriculum (COSMIC) project.* Paper presented at the Annual Meeting of the American Educational Research Association, New York.

National Council of Teachers of Mathematics. (1989). *Curriculum and evaluation standards for school mathematics.* Reston, VA: Author.

National Research Council. (2001). *Adding it up: Helping children learn mathematics.* J. Kilpatrick, J. Swafford, & B. Findell. (Eds). Mathematics Learning Study Committee, Center for Education, Division of Behavioral and Social Sciences and Education. Washington, DC: National Academy Press.

National Research Council. (2004). *On evaluating curricular effectiveness: Judging the quality of K-12 mathematics evaluations.* Committee for a Review of the Evaluation Data on the Effectiveness of NSF-Supported and Commercially Generated Mathematics Curriculum Materials. J. Confrey and V. Stohl (Eds.). Mathematical Sciences Education Board, Center for Education, Division of Behavioral and Social Sciences and Education. Washington, DC: National Academies Press.

No Child Left Behind (NCLB) Act of 2001, Pub. L. No. 107-110, § 115, Stat. 1425 (2002). Retrieved from http://www2.ed.gov/policy/elsec/leg/esea02/index.html

O'Donnell, C. L. (2008). Defining, conceptualizing, and measuring fidelity of implementation and its relationship to outcomes in K-12 curriculum intervention research. *Review of Educational Research, 78*(1), 33-84.

Pierce, R., Cassady, J., Adams, C., Speirs Neumeister, K. L., Dixon, F., & Cross, T. (2011). The effects of clustering and curriculum development on gifted students' math achievement. *Journal for the Education of the Gifted, 34,* 569-594.

Rectanus, C. (1994). *Math by all means: Geometry.* Sausalito, CA: Math Solutions.

Remillard, J. T. (2005). Examining key concepts in research on teachers' use of mathematics curricula. *Review of Educational Research, 75*(2), 211-246.

Remillard, J. T., & Heck, D. J. (2014). Conceptualizing the enacted curriculum in mathematics education. In D. R. Thompson & Z. Usiskin (Eds.), *Enacted mathematics curriculum: A conceptual framework and research needs* (pp. 121-148). Charlotte, NC: Information Age Publishing.

Remillard, J. T., Herbel-Eisenmann, B. A., & Lloyd, G. M. (Eds.). (2009). *Mathematics teachers at work: Connecting curriculum materials and classroom instruction.* New York, NY: Routledge.

Schoen, H. L, Cebulla, K. J., Finn, K. F., & Fi, C. (2003). Teacher variables that relate to student achievement when using a standards-based curriculum. *Journal for Research in Mathematics Education, 34*(3), 228-259.

Schmidt, W. H., Wolfe, R. G., & Kifer, E. (1992). The identification and description of student growth in mathematics achievement. In L. Burstein (Ed.), *The IEA study of mathematics III: Student growth and classroom processes* (pp. 59-99). Oxford: Pergamon Press.

Senk, S. L., & Thompson, D. R. (Eds.). (2003). *Standards-based school mathematics curriculum: What are they? What do students learn?* Mahwah, NJ: Lawrence Erlbaum.

Senk, S. L., Thompson, D. R., & Wernet, J. L. W. (2014). Curriculum and achievement in algebra 2: Influences of textbooks and teachers on students' learning about functions. In Y. Li & G. Lappan (Eds.), *Mathematics curriculum in school education* (pp. 515-540). New York, NY: Springer Dordrecht Heidelberg.

Stein, M. K., Remillard, J. T., & Smith, M. S. (2007). How curriculum influences student learning. In F. K. Lester (Ed.), *Second handbook of research on mathematics teaching and learning* (pp. 319-370). Charlotte, NC: Information Age Publishing.

Tarr, J. E., Chávez, Ó., Reys, R. E., & Reys, B. J. (2006). From the written to the enacted curricula: The intermediary role of middle school mathematics teachers in shaping students' opportunities to learn. *School Science and Mathematics, 106*(4), 191-201.

Tarr, J. E., Grouws, D. A., Chávez, Ó., & Soria, V. M. (2013). The effects of content organization and curriculum implementation on students' mathematics learning in second-year high school courses. *Journal for Research in Mathematics Education, 44*(4), 683-729.

Tarr, J. E., McKnaught, M. D., & Grouws, D. A. (2012). The development of multiple measures of curriculum implementation in secondary mathematics classrooms: Insights from a three-year curriculum evaluation study. In D. J. Heck, K. B. Chval, I. R. Weiss, & S. W. Ziebarth (Eds.), *Approaches to studying the enacted mathematics curriculum* (pp. 89-116). Charlotte, NC: Information Age Publishing.

Tarr, J. E., Reys, R. E., Reys, B. J., Chávez, Ó., Shih, J., & Osterlind, S. J. (2008). The impact of middle-grades mathematics curricula and the classroom learning environment on student achievement. *Journal for Research in Mathematics Education, 39*(3), 247-280.

Thompson, D. R., & Senk, S. L. (2001). The effects of curriculum on achievement in second-year algebra: The example of the University of Chicago School Mathematics Project. *Journal for Research in Mathematics Education, 32*, 58-84.

Thompson, D. R., & Senk, S. L. (2010). Myths about curriculum implementation. In B. Reys, R. Reys, & R. Rubenstein (Eds.), *Mathematics curriculum: Issues, trends, and future directions* (pp. 249-263). Reston, VA: National Council of Teachers of Mathematics.

Thompson, D. R., Senk, S. L., Witonsky, D., Usiskin, Z., & Kealey, G. (2001). *An evaluation of the second edition of UCSMP Advanced Algebra*. Chicago, IL: University of Chicago School Mathematics Project.

Thompson, D. R., Senk, S. L., & Yu, Y. (2012). *An evaluation of the third edition of UCSMP Transition Mathematics*. Chicago, IL: University of Chicago School Mathematics Project. Retrieved from http://ucsmp.uchicago.edu/research

Thompson, D. R., Witonsky, D., Senk, S. L., Usiskin, Z., & Kealey, G. (2003). *An evaluation of the second edition of UCSMP Geometry*. Chicago, IL: University of Chicago School Mathematics Project.

Valverde, G. A., Bianchi, L. J., Wolfe, R. G., Schmidt, W. H., & Houang, R. T. (2002). *According to the book: Using TIMSS to investigate the translation of policy into practice through the world of textbooks*. Dordrecht, Netherlands: Kluwer.

Weiss, I. R., Pasley, J. D., Smith, P. S., Banilower, E. R., & Heck, D. J. (2003). *Looking inside the classroom: A study of K-12 mathematics and science education in the United States*. Chapel Hill, NC: Horizon Research.

Ziebarth, S. W., Fonger, N. L., & Kratky, J. L. (2014). Instruments for studying the enacted mathematics curriculum. In D. R. Thompson & Z. Usiskin (Eds.), *Enacted mathematics curriculum: A conceptual framework and research needs* (pp. 97-120). Charlotte, NC: Information Age Publishing.

CHAPTER 4

TEACHERS' KNOWLEDGE AND THE ENACTED MATHEMATICS CURRICULUM

Ji-Won Son and Sharon L. Senk

Teachers' mathematical knowledge undoubtedly influences how mathematics curriculum is enacted during classroom instruction. This chapter considers what is known about teachers' knowledge of mathematics and curriculum, what differences in enactment occur based on differences in mathematical knowledge, and what more needs to be researched in this area.

INTRODUCTION

A strong school mathematics program is assumed to depend on a well-qualified teacher in every classroom and a challenging mathematics curriculum (Conference Board of the Mathematical Sciences [CBMS], 2012). This assumption raises questions about what a teacher should know to be considered well-qualified, and what knowledge is most helpful to enact a challenging mathematics curriculum. A desire to answer questions about the relation between teachers' knowledge and curriculum enactment led to the development of this chapter.

Shulman's (1986) conceptualization of knowledge for teaching, comprised of three components—pedagogical content knowledge (PCK), content knowledge (CK), and curriculum knowledge—has been widely accepted by the research community. CK has generally been defined as the conceptual knowledge of the subject matter being taught, which "refers to the amount and organization of knowledge per se in the mind of the teacher" (p. 9). PCK, in contrast, is the knowledge teachers need to make their subject matter comprehensible to students, which includes knowledge of the most appropriate instructional strategies and representations of ideas and the ability to anticipate and interpret students' understandings and misconceptions of subject matter. Curriculum knowledge is knowledge of the available instructional materials, as well as knowledge of the topics and the ways in which these were or will be addressed during the previous and subsequent years in schools (Rowland & Ruthven, 2011). In recent years, Usiskin (2000-01) has discussed "teachers' mathematics," which includes extensions of content being taught, concept analysis, and problem analysis. Ball, Thames, and Phelps (2008) introduced the phrase *mathematics knowledge for teaching* (MKT), which is arguably related to PCK for mathematics, and includes the use of representations to model mathematical concepts, evaluate representations and tasks in curriculum materials, provide students with explanations, and assess their solutions (Ball & Bass, 2003; Hill, Sleep, Lewis, & Ball, 2007).

As described in the chapter by Remillard and Heck (this volume), teachers' knowledge is assumed to influence how teachers teach and what students learn. A strong interdependence of teachers' knowledge and pedagogy has also been documented in numerous research reports (e.g., Fennema & Franke, 1992; Gamoran, 1994; Ma, 1999). In particular, a growing body of research provides evidence that a teacher's knowledge influences the implementation of curriculum materials that set ambitious goals for students (e.g., Ball & Cohen, 1996; Lloyd & Wilson, 1998; Son, in press; Spillane, 2000). This chapter was developed to investigate questions about the relation between teachers' knowledge and the enacted mathematics curriculum, specifically:

- What knowledge is used when curriculum is enacted?
- How is a teacher's use of curriculum materials mediated by his or her knowledge?
- In what tasks of enactment does teacher knowledge make itself most apparent?

In this chapter, we provide a synthesis of empirical research about the relation between teachers' knowledge and the enacted mathematics curriculum that was published in academic journals, books, and conference proceedings from 1989. We chose 1989 because it is the year the National Council of Teachers of Mathematics [NCTM] published its landmark *Curriculum and Evaluation Standards for School Mathematics*, which set the stage for ensuing decades of mathematics curriculum revision (Council of Chief State School Officers [CCSSO], 2010; NCTM, 2000). In the section that follows, we first describe what we mean by *knowledge* and *the enacted curriculum*. Then, we describe our search procedures, review the results of our search, and describe implications for educators and researchers. The purpose of the review is not to argue *whether* teachers' knowledge is related to teachers' mathematics curriculum enactment (it is), but *how* teachers' knowledge is related to the enacted curriculum and what can be done to support teachers' efforts to change their classroom practice.

METHODS

Assumptions and Definitions

Current educational reform efforts in mathematics education in the U.S. have been motivated by findings from international comparative studies about students' achievement and teaching methods (e.g., McKnight et al., 1987; Stigler & Hiebert, 1999); the recognition that the world and society into which students will graduate require greater ability to use mathematical tools (CCSSO, 2010; Heid, 1997); and by advances in pedagogy that emphasize building on students' prior knowledge, knowledge construction, and learning as a social activity (Fennema, Franke, Carpenter, & Carey, 1993; NCTM, 2000). In recent years, the development, dissemination, and implementation of curriculum materials aligned with the NCTM *Standards* documents (1989, 2000) have been used as a rationale for enhancing teachers' capacity to deliver high-quality mathematics instruction (CBMS, 2012; Remillard, 2005; Senk & Thompson, 2003).

Yet, changes in instruction do not occur simply because there are new curriculum materials in the classroom. As Remillard (2000) suggests, teachers' instructional practices evolve as a result of particular interactions between teachers and curriculum materials around specific subject matter and pedagogical content. In this chapter, teacher knowledge is assumed to encompass the aspects of knowledge described by Shulman (1986), but to exclude beliefs, although we recognize that teachers have frequently been found to treat their beliefs as knowledge (Thompson,

1992) and characteristics of teachers' belief systems have been linked to various components of knowledge. This conception about teacher knowledge may help us understand how CK, PCK, and curriculum knowledge are related to specific aspects of teaching.

By *enacted curriculum*, we mean what actually takes place in the classroom (e.g., the content and pedagogy actually delivered during instruction in the classroom) (Porter, 2004). Processes of constructing the enacted curriculum may include (a) planning, (b) predicting the sequence of classroom events, (c) interpreting students' actions and interactions, and (d) reflecting about events as or after they occur.

Selection Search Procedures

We used four criteria to select studies for review. First, for reasons described earlier, the search was limited to studies published between 1989 and 2012. Second, although there is much to be learned by looking at enacted curriculum within and across different content areas, we focused our review only on studies about mathematics education. Third, because curriculum materials and teaching practices vary so much around the world (Stigler & Hiebert, 1999; Valverde, Bianchi, Schmidt, McKnight, & Wolfe, 2002), we limited our search to studies conducted in the United States. Finally, we specifically sought studies that contained empirical evidence, either quantitative or qualitative, about the relation between teachers' knowledge and curriculum implementation. We excluded reports that gave prescriptions for practice based solely on intuition and experience as well as theory development articles in which no original data were reported. We also excluded studies that examined students' experiences with curriculum unless the study met the other criteria previously described.

We used a combination of manual and computer searches to conduct our review. We began by manually searching peer-reviewed research journals in mathematics education and general education that publish studies of mathematics instruction, including *Journal for Research in Mathematics Education*, *Journal of Mathematics Teacher Education*, *American Educational Research Journal*, *Elementary School Journal*, *Journal of Curriculum Studies*, and *Educational Studies in Mathematics*. To capture research that might not yet have been published, we used Google searches to locate references to related dissertations, books, and conference proceedings. The manual searches identified keywords that were later used in an ERIC search. The keywords in the first search, focused on enacted curriculum, were *mathematics* and either *implementation* or *enacted curriculum*. In the second search, focused on the role of teachers' knowledge on enacted curriculum, the keywords were *teacher knowledge, enacted curriculum, implemented curriculum, preservice teachers,* and *inservice teachers*. The database was expanded through a final manual search (i.e., references cited by studies caught in the initial search).

Coding Studies

Studies were coded for sample (preservice vs. inservice and grade levels of classrooms studied), methodology (qualitative vs. quantitative), results reported, and implications for teacher educators, curriculum developers, or policymakers. Whenever possible, we also coded for types of knowledge being explored (i.e., CK, PCK, & Curriculum Knowledge). However, not all researchers were explicit about which aspects of teachers' knowledge they were investigating, so in some cases such knowledge had to be inferred. We were particularly interested in the main findings from each study regarding how teachers' knowledge influences the nature of curriculum enactment.

RESULTS

Overall, we found more than 40 studies that addressed some aspect of the relation between teacher knowledge and mathematics instruction. However, in the majority of studies, the main focus was on *instructional practice*, not on the explicit relation between teacher knowledge and curriculum enactment. In only 8 of the studies was the relation between teacher knowledge and curriculum enactment addressed explicitly. Table 4.1 summarizes the nature of those eight studies. Only experienced teachers were involved in the studies described in Table 4.1.

We found few studies that explicitly addressed relations between teacher knowledge and the enacted curriculum and most involved very few teachers. In order to provide the fullest possible picture of relations between these variables, in this chapter we report on all studies uncovered by our search, not just those listed in Table 4.1.

We found two distinct, but related perspectives on teacher knowledge, curriculum materials, and enacted curriculum. In the first, the researchers focused on identifying the role of teacher knowledge during the enactment of curriculum materials. In the second, the researchers focused on the role of curriculum materials as tools for teacher learning.

Table 4.1. Characteristics of Studies That Explicitly Studied the Relation Between Teacher Knowledge and the Enacted Mathematics Curriculum

Author(s) and Year of Publication	N	Grade Levels(s)	Curriculum Materials Used
1. Anders (1995)	1	2-3	A script about addition with regrouping
2. Charalambous, Hill, & Mitchell (2012)	3	Middle grades	*Connected Mathematics Project*, integer subtraction
3. Hill & Charalambous (2012a)	2	7	*Connected Mathematics Project*, comparing ratios
4. Lewis & Blunk (2012)	2	8	*Connected Mathematics Project*, linear equations
5. Manouchehri (1998) and 6. Manouchehri & Goodman (1998)	66	Middle school	*Connected Mathematics Project, Mathematics in Context, Seeing and Thinking Mathematically, Sixth through Eighth Mathematics*
7. Sleep & Eskelson (2012)	2	6	*Connected Mathematics Project*, fractions
8. Stein, Baxter, & Leinhardt (1990)	1	5	Functions and graphing

Research on Role of Teacher Knowledge During Curriculum Enactment

Two themes emerged from the research on the role of teacher knowledge during curriculum enactment. The first is the effect of teacher knowledge on the enacted curriculum; the second is the effects of teacher knowledge, mediated by the use of curriculum materials, on their enactment of the curriculum.

Theme 1: Effects of Teacher Knowledge on the Enacted Curriculum

Early research investigating the effects of teachers' knowledge on their implementation of curricula can generally be classified into one of two groups: *deficit* or *affordance* approaches. In the former, the researchers draw linkages between a teacher's lack of mathematical understanding and patterns in his or her mathematics instruction (e.g., Ball & Cohen, 1996; Thompson, 1992; Wilson, 1990); in the latter approach, the researchers highlight the affordances strong mathematics understandings create for classroom instruction (e.g., Hill, Schilling, & Ball, 2004).

In the deficit studies, the researchers observed significant mathematical errors or imprecision during classroom instruction, such as inappropriate metaphors for mathematical procedures (Heaton, 1992), incomplete definitions (Stein, Baxter, & Leinhardt, 1990), difficulties presenting mathematical materials clearly and correctly (Charalambous, Hill, & Mitchell, 2012), or other mathematical mistakes (Putnam, Heaton, Prawat, & Remillard, 1992). Other studies report that teachers lack the disciplinary knowledge required to make full use of rich problems in the curriculum materials as designed by the curriculum developers (Henningsen & Stein, 1997; Lloyd & Wilson, 1998; Spillane, 2009). For example, Stein, Baxter and Leinhardt (1990) document one teacher's limited knowledge of functions and observe that he offered students a definition that was missing several key elements. The teacher also presented materials in a way that did not provide a foundation for future development of the topic. Similarly, Putnam et al. (1992) demonstrated that teachers with limited CK tended to make curriculum adaptations focusing on rules and procedures. Wilson (1990), Gamoran (1994), and Herbel-Eisenmann and Otten (2011) also describe teachers whose knowledge of mathematics and teaching limit their abilities to implement curriculum materials in ways intended by the curriculum developers or policymakers. These deficit studies imply that low mathematical knowledge leads to mathematical errors and poor mathematical choices in the enacted curriculum, even when teachers are using curriculum materials that provide opportunities for more cognitively demanding work.

In contrast, affordance studies focused on the practice of teachers engaged in using new curriculum materials and the strong positive association of teacher knowledge and the enacted curriculum (Hill & Charalambous, 2012b; Lloyd, 2002; Lloyd & Wilson, 1998; Manouchehri, 1998; Manouchehri & Goodman, 1998, 2000). Researchers in these studies examined what teachers with higher knowledge can do that others cannot, focusing on both CK and PCK in the use of curriculum materials (e.g., Anders, 1995; Lloyd, 2002; Manouchehri & Goodman, 2000). Affordance studies demonstrate that teachers with more knowledge exhibit greater "richness of the mathematics available for the learners" (Fennema & Franke, 1992, pp. 150-151). Such studies also show that a teacher with stronger CK and PCK supports more complete implementation of curriculum materials than a teacher with weaker mathematical knowledge. For example, Anders (1995) reported that teacher knowledge influences all four processes of enacting curriculum—(1) planning lessons, (2) predicting the sequence of classroom events, (3) interpreting students' actions and interactions, and (4) reflecting about events as or after they occurred—but particularly the second and third processes. She demonstrated that Ms. G's knowledge about students' difficulties or misconceptions (PCK) affected the selection and use of examples and strategies (i.e., teaching addition with regrouping using a game with base-10 blocks). During the classroom event in which each task was introduced and enacted, the teacher's CK and PCK influenced the interpretation of students' responses and behaviors to make decisions about how to proceed through activities.

Similarly, Lloyd and Wilson (1998), who examined the effects of a teacher's conceptions of functions on his implementation of a new high school curriculum, showed that the teacher's strong graphical and covariation-centered concept image played a crucial supporting role in the teacher's adaptation of the *Core-Plus* approach to functions. This approach contributed to instructional practices that encouraged students to utilize a variety of representations and connections among them to investigate real-world occurrences of different families of functions.

In the only study we found with a large sample of teachers ($n = 66$), Manouchehri and Goodman (1998) observed substantial differences in how middle school teachers used new curriculum materials over a period of 2 years. Teachers varied in (a) time they spent using the materials, (b) expectations from the students, and (c) amount of effort invested in building the type of classroom culture conducive to students' productive use of the materials. Teachers' content background and confidence in their knowledge of mathematics influenced whether the materials were consistently used in classrooms. The teachers with weaker mathematical knowledge avoided those mathematics units they did not feel comfortable teaching or those they did not view as mathematically significant. The authors reported that what teachers knew about mathematics content and innovative pedagogical practices and their personal theories about learning and teaching mathematics were the greatest influences on how teachers valued and implemented their programs. When the teachers did not have immediate answers to the questions students generated, they either dismissed or ignored them. The consequence of this practice was detrimental to both teachers and students.

Several more recent studies (e.g., Choppin, 2011; Drake, 2002, 2006) have examined the relationship between teacher curriculum knowledge and curriculum enactment, in particular, on the process of curriculum adaptation. Adaptation is a central process in teachers' use of curriculum materials, because curriculum is often not used blindly without modification (Ben-Peretz, 1990, 2011). Choppin (2011) examined three teachers and their use of curriculum materials, focusing on (1) how the teachers drew on the materials, (2) what they understood about the curriculum resources, and (3) how they connected their use of the materials to their observations of student thinking. He showed that similarities existed across the teachers, particularly with respect to their goals and how they read and followed recommendations in the teacher resource materials.

However, differences existed in how they revised tasks within the curriculum materials while enacting them in their classroom in response to what they observed about student thinking. The teacher who most intensively observed and understood student thinking made connections between her interpretations of students' strategies and her use of the curriculum resources, allowing her to design learned adaptations. Learned adaptations are adaptations designed based on what teachers have learned from prior enactments. They involve understanding how curriculum resources can be used to design instruction with respect to particular outcomes, such as fostering opportunities for students to develop conceptual understanding through active engagement in mathematical processes. Therefore, learned adaptations require both an understanding of the curriculum developers' design rationale and empirically developed knowledge of how that rationale has played out in practice.

Huntley and Chval (2010) investigated how teachers used their assigned textbooks, basing their investigation on teachers' self reports. Data came from two separate middle-grades research projects. One project studied the implementation of three comprehensive middle-grades mathematics curricula developed with funding from the National Science Foundation, called "NSF-funded textbooks"; the other project studied teachers' use of district-adopted textbooks that were called "publisher-generated textbooks."[1] Huntley and Chval report teachers make decisions to supplement or modify their textbooks for a variety of reasons, depending on the type of curriculum materials they use. Approximately 95% of the teachers who were using NSF-funded textbooks reported going from one page to the next in sequential fashion. These teachers expressed two reasons for this approach: the nature of the NSF-funded curriculum; and the authority of textbooks. They believed that, in the NSF-funded textbooks, the mathematical ideas build on one another. These teachers also expressed respect for the textbook authors' judgment.

In contrast, 65% of the teachers using publisher-generated textbooks stated that they followed the textbook sequentially. Among the 35% who did not, teachers reported that the textbook did not seem to be set up in an orderly system for the teachers; therefore, they switched the order and ignored the sequence in the text. The most common reason for not following the textbook's order was demands of their state's test. In other words, teachers wanted to make sure that certain mathematical content was presented prior to the time their students took the state assessment.

Both sets of teachers reported omitting some material from their textbooks. And, both groups of teachers reported perceiving that they had insufficient time available. Teachers of NSF-funded textbooks reported perceiving that problems were redundant; teachers using publisher-generated curricula reported that students did not need to know the content or that other content was more important (e.g., problem-solving was omitted because teachers believe students need to focus on skills). These findings highlight various factors other than content knowledge that influence teachers' curriculum enactment.

Theme 2: Effects of Teacher Knowledge, Mediated by the Use of Curriculum Materials, on Their Curriculum Enactment

Recently, Hill and her colleagues conducted a set of case studies to investigate the relations among teacher knowledge, curriculum materials, and quality of instruction (Charalambous, Hill, & Mitchell, 2012; Hill & Charalambous, 2012a, 2012b; Lewis & Blunk, 2012; Sleep & Eskelson, 2012). The studies show that the mathematical quality of instruction in classrooms is a function of differences in teacher knowledge and that curriculum materials can mediate the knowledge of teachers with low or weaker mathematical knowledge for teaching. Charalambous, Hill, and Mitchell (2012), for example, explored how limitations in teacher knowledge and the curriculum contribute to instructional quality. This study examined the contribution of mathematical knowledge for teaching (MKT) and curriculum materials to the implementation of lessons on integer subtraction. They investigated the instruction of three teachers with differing MKT levels using two editions of the same set of curriculum materials that provided different levels of support. Charalambous, Hill, and Mitchell reported that MKT relates positively to teachers' use of representations, provision of explanations, precision in language and notation, and ability to capitalize on students' contributions to move the mathematics forward in a goal-directed manner. In particular, arguing that curriculum materials set the stage for attending to the meaning of integer subtraction, they reported that less educative curriculum materials, coupled with low levels of MKT, lead to problematic instruction while educative materials helped low-MKT teachers provide adequate instruction.

Sleep and Eskelson (2012) compared the enactment of a problem about fractions taught by two teachers with differing levels of MKT. MKT seemed to support teachers' precise use of mathematical language and helped to prevent errors. Additionally, the curriculum materials provided rich opportunities for mathematical work. Lewis and Blunk (2012) reported similar findings from a study of lessons about linear equations taught by two teachers with different levels of MKT. Although the two teachers were teaching from the same curriculum materials, the teacher with higher MKT had more complete and concise ways to describe key concepts, had multiple ways to represent ideas about linear equations, could move among different mathematical expressions of linear relationships, and gave students a larger role in articulating the mathematical ideas of the lesson. They also reported that the lower-MKT teacher stayed on topic, and defined concepts in reasonably accurate ways when he followed the curriculum materials closely. However, he also made minor mathematical errors.

Hill and Charalambous (2012a) also documented ways that MKT and curriculum materials appeared to contribute to the enactment of a seventh grade *Connected Mathematics Project* lesson on comparing ratios by two teachers with different levels of MKT. Although novel tasks contained in the curriculum materials may present difficulties to teachers initially, they reported that these tasks also lay the groundwork for in-depth student problem-solving experiences. Only the higher-MKT teacher was successfully able to meet the challenge by following the curriculum closely. These case studies suggest that lower levels of mathematical knowledge can sometimes be mediated by educative curriculum materials.

Research on the Role of Curriculum Materials on Teacher Learning

While the aforementioned studies examined how teacher knowledge or knowledge mediated by educative curriculum materials influences teachers' enactment of curriculum, other scholars have investigated the opposite direction, that is, how curriculum materials influence teacher knowledge. This group of researchers investigates the use of innovative curriculum materials as a tool for learning mathematics by both preservice and inservice teachers.

A growing body of research (Empson & Junk, 2004; Papick, Beem, Reys, & Reys, 1999; Remillard, 2000, 2009; Remillard & Bryan, 2004; Sowder, Philipp, Armstrong, & Schappelle, 1998; Wilson & Lloyd, 2000) explored both the challenges to teachers of using novel curriculum materials in mathematics classrooms and the potential role curriculum materials can play in supporting teachers' knowledge and efforts to develop expertise with new forms of mathematics instruction. Remillard (2000), for example, examined whether and how teachers' use of a new textbook contributes to their learning and related changes in their mathematics teaching. From analyses across cases of two 4th grade teachers, she reported that activities that included analyzing student thinking and mathematical tasks and making decisions about how to proceed were likely to foster teacher learning.

Sherin (2002) investigated how five high school teachers implemented new instructional materials. The study found that the teachers used their mathematics content knowledge in three different ways: (1) to alter the mathematical aspects of the lessons to fit their own prior understandings of the mathematics content rather than following the intent of the curriculum developers; (2) to teach aspects of the lessons in ways that differed from, and extended, their own prior mathematics content knowledge; and (3) to use new content knowledge developed as a product of implementing the lessons to direct the course of the lesson in ways that differed from what was explicitly intended by the developers.

Some researchers also have examined the influence of *Standards*-based curricula on preservice teachers' learning (Lloyd & Frykholm, 2000; Papick, Beem, Reys, & Reys, 1999; Stump, Bishop, & Britton, 2003). Here, the term *enacted curriculum* is different from previous studies of teacher learning with inservice teachers. In this line of studies, the term *enacted curriculum* describes the particular learning contexts "jointly constructed by teachers, students, and materials," typically in a course on methods of teaching mathematics (Ball & Cohen, 1996, p. 7). Lloyd (2002) suggested that *Standards*-based curriculum materials may offer fruitful resources with which preservice teachers can learn about mathematics and pedagogy. Because such curriculum materials provide novel ideas about both mathematics and pedagogy, they provide opportunities for work in which preservice teachers can revisit their knowledge of mathematics and their beliefs about pedagogy. However, these studies did not provide a direct relationship between teacher knowledge and curriculum materials, and they (i.e., Lloyd & Frykholm, 2000) reported that textbook type (i.e., *Standards*-based or more traditional commercially-published text) is not an indicator for improving preservice teachers' knowledge.

Spielman and Lloyd (2004) examined the impact of instructional materials on preservice teachers' learning of mathematics. They arranged for two classes of mathematics for elementary school teachers to be taught by the same instructor using different curriculum materials, one emphasizing rules and procedures and the other emphasizing reform-oriented approaches. Although this study found significant differences between the two groups with respect to their conceptions or beliefs about mathematics instruction, they found no difference on a test of mathematics content knowledge that included mathematics content such as place value, fractions, and probability.

In another study, Lloyd (2009) described ways that five preservice teachers viewed and interacted with the rhetorical components of innovative school mathematics curriculum materials used in a mathematics course for future elementary teachers. The preservice teachers' comments reflected general agreement that the innovative curriculum materials contained fewer narrative elements and worked examples, as well as more (and different) exercises and question sets and activity elements, than the mathematics textbooks to which the teachers were accustomed. However, according to Lloyd, variation emerged when considering the ways in which the teachers interacted with the materials for their learning of mathematics. Although each preservice teacher considered that use of the curriculum materials improved her mathematical understandings in significant ways, some accepted and even embraced changes to the teaching–learning process that accompanied use of the curriculum materials; in contrast, others experienced discomfort and frustration at times.

SUMMARY

Teachers' knowledge and curriculum materials are two key components that contribute to the quality of school mathematics instruction (CBMS, 2012; Stein, Remillard, & Smith, 2007). Consequently, over the last quarter century, researchers have focused on both teachers' knowledge and the use of curriculum materials as variables of interest.

We set out to answer three questions: (a) What knowledge is used when curriculum is enacted? (b) How is a teacher's use of curriculum materials mediated by his or her knowledge? (c) In what tasks of enactment does teacher knowledge make itself most apparent? Thus far, some patterns have emerged to answer our first two questions.

We begin by considering the interdependence between teacher knowledge and curriculum enactment. Many studies showed that teacher knowledge is a key contributor to curriculum enactment. All three types of knowledge (i.e., content knowledge, pedagogical content knowledge, and curriculum knowledge) have a bearing on teachers' use of instructional materials. Of the studies using Shulman's conceptualization, those involving CK were the most common; those involving curriculum knowledge were the least common. Content knowledge influences mathematics teachers' instructional decisions when using materials. Teachers with less mathematics knowledge tended to focus on algorithms rather than on underlying mathematics concepts, and they often made mathematical errors when presenting concepts. In contrast, teachers who understood multiple representations of mathematics concepts were able to use these representations to further students' understanding. More knowledgeable teachers with PCK tended to present problems in contexts that were familiar to the students and to link problems to what students had already learned. They were also more likely to approach students' questions mathematically and to solve problems collaboratively with students, compared to less knowledgeable teachers who tended to look up correct answers in response to students' questions. When planning lessons on familiar content, teachers with PCK and curriculum knowledge had a sense of how to build a storyline by presenting concepts in a logical sequence. In areas in which they were unfamiliar with the content, they were aware of the need for appropriate sequencing, but were unable to identify the key concepts.

Hill and Charalambous (2012b) observed similar relations between teacher knowledge and curriculum enactment. In particular, they suggested the following ways in which strong mathematical knowledge for teaching may afford specific advantages in teaching mathematics in the enactment of curriculum:

- Use of mathematical language in explanations;
- Making connections across ideas and representations, and using multiple methods;
- Sequencing and linking lesson tasks and activities in order to help students gradually build the core mathematical ideas of the lesson.

According to Hill and Charalambous (2012b), teachers with high MKT tend to use mathematical language both densely and precisely. During times of whole-class instruction, these teachers infuse their talk with mathematical terms, resulting in descriptions and explanation rife with technical language. These teachers also tend to articulate accurate and helpful explanations for the mathematical ideas in the lesson while drawing explicit connections among and between important mathematical ideas and different representations. At the same time, high MKT teachers tend to weave the lesson activities together and retain the focus on key mathematical ideas while they quickly understand and are then able to capitalize on student ideas to move the lesson forward toward larger mathematical goals.

Several research studies reported that teachers' beliefs about curriculum materials played a role as a mediator between their knowledge and curriculum enactment. A teacher who is oriented toward student exploration and who has more knowledge for teaching might be better able to launch a task than a teacher with less knowledge. That is, teachers' knowledge of mathematics and pedagogy translates into practice through the filter of their beliefs (Cooney, 1994). Therefore, teachers' knowledge is not linearly or consistently related to enacting the mathematics curriculum.

Several studies have shown that curriculum materials contribute to the development of teacher knowledge, teacher learning, and curriculum enactment. Charalambous, Hill, and Mitchell (2012) report that the curriculum materials also relate to the quality of instruction, especially when they are sufficiently supportive for teachers. In particular, they document that, when providing adequate levels of support, curriculum materials can enable teachers to provide instruction that supports meaning-making, help them avoid mathematical errors and notational linguistic imprecision, and encourage them to engage students in cognitively demanding activities. Even teachers with weaker mathematical knowledge for teaching could provide adequate instruction by staying close to their curriculum materials. Thus, the support designed into the curriculum materials matters in conjunction with teacher knowledge and beliefs. In particular, when the materials clarify the rationale behind proposed tasks and activities and make key mathematical ideas transparent to teachers, they are more likely to have a positive impact on instructional quality (Son, in press). In contrast, when the materials are not sufficiently supportive, or are judged to increase the difficulty of lesson enactment through their design, low quality enactment tends to occur in classrooms.

LIMITATIONS

Although previous studies suggest relations between teacher knowledge and the enacted curriculum, several limitations of this review can be noted. First, our review is not comprehensive. Searching additional journals might uncover additional studies that have explicitly investigated relations between teacher knowledge and the enacted curriculum.

Second, not all research studies we found used the same conceptions of teachers' knowledge. Some studies distinguish CK, PCK, and curriculum knowledge, the aspects of knowledge for teaching distinguished by Shulman (1986); others use mathematical knowledge for teaching as measured by instruments developed by Ball, Hill, and others (Ball et al., 2008; Hill et al., 2008). On the one hand, the diversity of conceptions of knowledge leads to rich descriptions; on the other hand, it limits generalizability. We agree with the call from the National Research Council [NRC] (2010) that researchers should "clarify what is meant by teacher knowledge and how that construct can best be measured, and how content knowledge interacts with knowledge of the pedagogical application of that knowledge" (p. 197). Specifically, individual researchers should indicate what definition of CK, PCK, curriculum knowledge, and MKT their study uses, even if the entire community may not agree on definitions of these terms.

Third, most of the research examining relations between teachers' knowledge and curriculum enactment has been conducted with very small samples and used qualitative methods. Although the descriptions in these studies are often very rich, one must be careful not to overgeneralize. To improve generalizability, larger samples and more use of experimental or quasi-experimental designs should be employed.

Fourth, almost all of the studies we reviewed, including all that are listed in Table 4.1, were conducted with elementary or middle school teachers. Most elementary school teachers and some middle school mathematics teachers are not mathematics specialists. Thus, it is not surprising that many studies involved teachers with low levels of CK or MKT. In contrast, most high school mathematics teachers have degrees in mathematics or mathematics education (Banilower, Smith, Weiss, Malzahn, Campbell, & Weiss, 2013). More research about the relation between teachers' knowledge and curriculum enactment needs to be conducted with samples of high school teachers to understand how the patterns described in this review for non-mathematics specialists extend to specialists.

WHERE DO WE GO FROM HERE?

As noted by Remillard and Heck (this volume), the enacted curriculum is influenced by many factors, including teachers' knowledge. As discussed in this review, the curriculum materials may also influence a teacher's knowledge; that is, teachers learn from curriculum materials. More research also needs to be done in the area of the effect of curriculum materials on teacher learning. The field lacks detailed understanding regarding how curriculum materials help inservice and preservice teachers enhance their understanding for teaching. We recommend a series of questions to guide discussions related to teacher knowledge, curriculum materials and ultimately decision-making in relation to other factors:

- What factors relate to curriculum enactment?
- In what tasks of enactment does teacher knowledge make itself most apparent?
- How is a teacher's use of curriculum materials mediated by his or her knowledge and other factors?

By the very nature of the profession, teachers will continue to supplement, omit problems or sections, and change the order of lessons presented in textbooks. Attending to the aforementioned research questions could yield useful recommendations for policy and teacher education, especially given recent studies that suggest that simply injecting ambitious curricula into the instructional system does not guarantee high-quality instruction. By concurrently attending to teachers' knowledge and their use of curriculum materials, we hope that researchers could respond to recent recommendations for exploring how teacher knowledge, in conjunction with other systemic resources and contextual factors, contribute to instructional quality.

NOTE

1. The NSF-funded textbooks were *Connected Mathematics, Mathematics in Context,* and *Math Thematics;* the publisher-generated textbooks were produced by Glencoe, Saxon, Prentice Hall, Houghton-Mifflin, Southwestern, Harcourt Brace, and Addison-Wesley.

REFERENCES

Anders, D. (1995). A teacher's knowledge as classroom script for mathematics instruction. *Elementary School Journal, 95*(4), 311-324.

Ball, D. L., & Bass, H. (2003). Making mathematics reasonable in school. In J. Kilpatrick, W. G. Martin, & D. Schifter (Eds.), *A research companion to principles and standards for school mathematics* (pp. 27-44). Reston, VA: National Council of Teachers of Mathematics.

Ball, D. L., & Cohen, D. K. (1996). Reform by the book: What is: Or might be: The role of curriculum materials in teacher learning and instructional reform? *Educational Researcher, 25*(9), 6-8.

Ball, D. L., Thames, M. H., & Phelps, G. (2008). Content knowledge for teaching: What makes it special? *Journal of Teacher Education, 59*(5), 389-407.

Banilower, E. R., Smith, P. S., Weiss, I. R., Malzahn, K. M., Campbell, K. M., & Weiss, A. M. (2013). *Report on the 2012 national survey of science and mathematics education*. Chapel Hill, NC: Horizon Research.

Ben-Peretz, M. (1990). *The teacher-curriculum encounter: Freeing teachers from the tyranny of texts*. Albany: State University of New York Press.

Ben-Peretz, M. (2011). Teacher knowledge: What is it? How do we uncover it? What are its implications for schooling? *Teaching and Teacher Education, 27,* 3-9.

Charalambous, C., Hill, H., & Mitchell, R. (2012). Two negatives don't always make a positive: Exploring how limitations in teacher knowledge and the curriculum contribute to instructional quality. *Journal of Curriculum Studies, 44*(4), 489-513.

Choppin, J. (2011). Learned adaptations: Teachers' understanding and use of curriculum resources. *Journal of Mathematics Teacher Education, 14,* 331-353.

Conference Board of the Mathematical Sciences. (2012). *The mathematical education of teachers II*. Providence, RI: American Mathematical Society and Mathematical Association of America.

Cooney, T. (1994). Research and teacher education: In search for common ground. *Journal for Research in Mathematics Education, 25*(6), 608-636.

Council of Chief State School Officers. (2010). *Common Core state standards for mathematics*. (2010). (Retrieved November 10, 2010 from http://www.corestandards.org/assets/CCSSI_Math%20Standards.pdf)

Drake, C. (2002). Experience counts: Career stage and teachers' responses to mathematics education reform. *Educational Policy, 16*(2), 311-337.

Drake, C. (2006). Turning points: Using teachers' mathematics life stories to understand the implementation of mathematics education reform. *Journal of Mathematics Teacher Education, 9,* 579-608.

Empson, S. B., & Junk, D. L. (2004). Teachers' knowledge of children's mathematics after implementing a student-centered curriculum. *Journal of Mathematics Teacher Education, 7*(2), 121-144.

Fennema, E., & Franke, M. L. (1992). Teachers' knowledge and its impact. In D. A. Grouws (Ed.), *Handbook of research on mathematics teaching and learning: A project of the National Council of Teachers of Mathematics* (pp. 147-164). New York, NY: Macmillan.

Fennema, E., Franke, M. L., Carpenter, T. P., & Carey, D. A. (1993). Using children's mathematical knowledge in instruction. *American Educational Research Journal, 30*(3), 555-583.

Gamoran, A. (1994). Schooling and achievement: Additive versus interactive models. In I. Westbury, C. A. Ethington, L. A. Sosniak, & D. P. Baker (Eds.), *In search of a more effective mathematics education* (pp. 273-292). Norwood, NJ: Ablex Publishing Corporation.

Heaton, R. M. (1992). Who is minding the mathematics content? A case study of a fifth-grade teacher. *The Elementary School Journal, 93*(2), 153-162.

Heid, M. K. (1997). The technological revolution and the reform of school mathematics. *American Journal of Education, 106*(1), 5-61.

Henningsen, M., & Stein, M. K. (1997). Mathematical tasks and student cognition: Classroom-based factors that support and inhibit high-level mathematical thinking and reasoning. *Journal for Research in Mathematics Education, 28*(5), 524-549.

Herbel-Eisenmann, B. A., & Otten, S. (2011). Mapping mathematics in classroom discourse. *Journal for Research in Mathematics Education, 42*, 451-485.

Hill, H., & Charalambous, C. (2012a). Teaching (un)connected mathematics: Two teachers' enactment of the pizza problem. *Journal of Curriculum Studies, 44*(4), 467-487.

Hill, H., & Charalambous, C. (2012b). Teacher knowledge, curriculum materials, and quality of instruction: Lessons learned and open issues. *Journal of Curriculum Studies, 44*(4), 559-576.

Hill, H. C., Blunk, M., Charalambous, C. Y., Lewis, J., Phelps, G. C., Sleep, L., & Ball, D. L. (2008). Mathematical knowledge for teaching and the mathematical quality of instruction: An exploratory study. *Cognition and Instruction, 26*, 430-511.

Hill, H. C., Schilling, S. G., & Ball, D. L. (2004). Developing measures of teachers' mathematics knowledge for teaching. *Elementary School Journal, 105*, 11-30.

Hill, H. C., Sleep, L., Lewis, J. M., & Ball, D. L. (2007). Assessing teachers' mathematical knowledge: What knowledge matters and what evidence counts? In F. K. Lester (Ed.), *Second handbook of research on mathematics teaching and learning* (pp. 111-155). Charlotte, NC: Information Age.

Huntley, M., & Chval, K. (2010). Teachers' perspectives on fidelity of implementation to textbooks. In B. Reys, R. Reys, & R. Rubenstein (Eds.), *The K–12 mathematics curriculum: Issues, trends, and future directions* (pp. 289-304). Reston, VA: National Council of Teachers of Mathematics.

Lewis, J., & Blunk, M. (2012). Reading between the lines: Teaching linear algebra. *Journal of Curriculum Studies, 44*(4), 515-536.

Lloyd, G. M. (2002). Reform-oriented curriculum implementation as a context for teacher development: An illustration from one mathematics teacher's experience. *The Professional Educator, 24*(2), 51-61.

Lloyd, G. M. (2009). School mathematics curriculum materials for teachers' learning: Future elementary teachers' interactions with curriculum materials in a mathematics course in the United States. *ZDM-The International Journal on Mathematics Education, 41*, 763-775.

Lloyd, G. M., & Frykholm, J. A. (2000). How innovative middle school mathematics materials can change prospective elementary teachers' conceptions. *Education, 21*, 575-580.

Lloyd, G. M., & Wilson, M. S. (1998). Supporting innovation: The impact of a teacher's conceptions of functions on his implementation of a reform curriculum. *Journal for Research in Mathematics Education, 29*(3), 248-274.

Ma, L. (1999). *Knowing and teaching elementary mathematics: Teacher's understanding of fundamental mathematics in China and the United States.* Mahwah, NJ: Lawrence Erlbaum Associates.

Manouchehri, A. (1998). Mathematics curriculum reform and teachers: What are the dilemmas? *Journal of Teacher Education, 49*(4), 276-286.

Manouchehri, A., & Goodman, T. (1998). Mathematics curriculum reform and teachers: Understanding the connections. *Journal of Educational Research, 92*(1), 27-41.

Manouchehri, A., & Goodman, T. (2000). Implementing mathematics reform: The challenge within. *Educational Studies in Mathematics, 42*, 1-34.

McKnight, C. C., Crosswhite, F. J., Dossey, J. A., Kifer, E., Swafford, J. O., Travers, K. J., & Cooney, T. J. (1987). *The underachieving curriculum: Assessing U.S. school mathematics from an international perspective.* Champaign, IL: Stipes.

National Council of Teachers of Mathematics. (1989). *Curriculum and evaluation standards for school mathematics.* Reston, VA: Author.

National Council of Teachers of Mathematics. (2000). *Principles and standards for school mathematics.* Reston, VA: Author.

National Research Council. (2010). *Preparing teachers: Building evidence for sound policy.* Committee on the Study of Teacher Preparation Programs in the United States, Center for Education. Division of Behavioral and Social Sciences and Education. Washington, DC: The National Academies Press.

Papick, J. J., Beem, J. K., Reys, B. J., & Reys, R. E. (1999). Impact of the Missouri middle mathematics project on the preparation of prospective middle school teachers. *Journal of Mathematics Teacher Education, 2*(3), 301-310.

Porter, A. (2004). *Curriculum assessment (Additional SCALE Research Publications and Products: Goals 1, 2, and 4).* Nashville, TN: Vanderbilt University.

Putnam, R. T., Heaton, R. M., Prawat, R. S., & Remillard, J. (1992). Teaching mathematics for understanding: Discussing case studies of four fifth-grade teachers. *The Elementary School Journal, 93*(2), 213-228.

Remillard, J. T. (2000). Can curriculum materials support teachers' learning? *Elementary School Journal, 100*(4), 331-350.

Remillard, J. T. (2005). Examining key concepts in research on teachers' use of mathematics curricula. *Review of Educational Research, 75*(2), 211-246.

Remillard, J. T. (2009). Considering what we know about the relationship between teachers and curriculum materials. In J. T. Remillard, B. A. Herbel-Eisenmann, & G. M. Lloyd (Eds.), *Mathematics teachers at work: Connecting curriculum materials and classroom instruction* (pp. 85-92). New York, NY: Routledge.

Remillard, J. T., & Bryans, M. B. (2004). Teachers' orientations toward mathematics curriculum materials: Implications for teacher learning. *Journal for Research in Mathematics Education, 35*(5), 352-388.

Remillard, J. T., & Heck, D. J. (2014). Conceptualizing the enacted curriculum in mathematics education. In D. R. Thompson & Z. Usiskin (Eds.), *Enacted mathematics curriculum: A conceptual framework and research needs* (pp. 75-95). Charlotte, NC: Information Age Publishing.

Rowland, T., & Ruthven, K. (2011). (Eds.). *Mathematical knowledge in teaching.* London and New York: Springer.

Senk, S. L., & Thompson, D. R. (2003). *Standards-based school mathematics: What are they? What do students learn?* Mahwah, NJ: Erlbaum.

Sherin, M. G. (2002). When teaching becomes learning. *Cognition and Instruction, 20*(2), 119-150.

Shulman, L. S. (1986). Those who understand: Knowledge growth in teaching. *Educational Researcher, 14,* 4-14.

Sleep, L., & Eskelson, S. L. (2012). MKT and curriculum materials are only part of the story: Insights from a lesson on fractions. *Journal of Curriculum Studies, 44*(4), 537-558.

Son, J. (in press). Factors that influence mathematics teachers' enacted curriculum: Cognitive demands of mathematical problems and teacher-guided questions. *Journal of Curriculum Studies.*

Sowder, J. T., Philipp, R. A., Armstrong, B. E., & Schappelle, B. P. (1998). *Middle-grade teachers' mathematical knowledge and its relationship to instruction.* Albany, NY: State University of New York Press.

Spielman, L. J., & Lloyd, G. M. (2004). The impact of enacted mathematics curriculum models on prospective elementary teachers' course perceptions and beliefs. *School Science and Mathematics, 104*(1), 32-44 [reprinted online March, 2010].

Spillane, J. P. (2000). A fifth-grade teacher's reconstruction of mathematics and literacy teaching: Exploring interactions among identity, learning, and subject matter. *The Elementary School Journal, 100*(4), 307-330.

Spillane, J. P. (2009). Leading and managing instruction: Adopting a diagnostic and design mindset. *Voices in Urban Education: Leadership in Smart Systems, 25*(Fall), 16-25.

Stein, M. K., Baxter, J. A., & Leinhardt, G. (1990). Subject-matter knowledge and elementary instruction: A case from functions and graphing. *American Educational Research Journal, 27*(4), 639-663.

Stein, M., Remillard, J., & Smith, M. S. (2007). How curriculum influences student learning. In F. K. Lester (Ed.), *Second handbook of research on mathematics teaching and learning* (pp. 319-369). Charlotte, NC: Information Age Publishing.

Stigler, J., & Hiebert, J. (1999). *The teaching gap: Best ideas from the world's teachers for improving education in the classroom.* New York, NY: Free Press.

Stump, S., Bishop, J., & Britton, B. (2003). Building a vision of algebra for preservice teachers. *Teaching Children Mathematics*, *10*(3), 180-186.

Thompson, A. G. (1992). Teachers' beliefs and conceptions: A synthesis of the research. In D. A. Grouws (Ed.), *Handbook of research on mathematics teaching and learning* (pp. 127-146). New York, NY: Macmillan.

Usiskin, Z. (Winter 2000-01). Teachers' mathematics: A collection of content deserving to be a field. *UCSMP Newsletter*, *No. 28*, 5-10. (Reprinted in *The Mathematics Educator* (Singapore). (2001). *Volume 6*(1), 85-98.)

Valverde, G. A., Bianchi, L. J., Schmidt, W. H., McKnight, C. C., & Wolfe, R. G. (2002). *According to the book: Using TIMSS to investigate the translation of policy into practice in the world of textbooks*. Dordrecht, Netherlands: Kluwer.

Wilson, S. M. (1990). A conflict of interests: The case of Mark Black. *Educational Evaluation and Policy Analysis*, *12*, 293-310.

Wilson, M. R., & Lloyd, G. M. (2000). The challenge to share mathematical authority with students: High school teachers reforming classroom roles. *Journal of Curriculum and Supervision*, *15*, 146-169.

CHAPTER 5

INSTRUMENTS FOR STUDYING THE ENACTED MATHEMATICS CURRICULUM

Steven W. Ziebarth, Nicole L. Fonger, and James L. Kratky

This chapter discusses the variety of approaches that researchers have used to study the enacted curriculum by examining some of the instruments that have been developed and reported through research journals and other dissemination outlets. The focus of the chapter is on a research database of instruments developed and maintained over the past eight years by the Center for the Study of Mathematics Curriculum. The chapter begins with a brief description and history of the database's development, followed by short discussions of various types of instruments within prominent categories of the database. It concludes with some observations focused on where gaps yet exist for the possible development of new instruments to move curriculum research forward.

INSTRUMENT DATABASE BACKGROUND

In 2005, as part of the early work of the Center for the Study of Mathematics Curriculum[1] (CSMC), faculty and graduate fellows[2] from the participating institutions began developing two databases for use by researchers and others interested in research on mathematics curriculum. Both databases are now among four currently (2013) active on the CSMC website: *Mathematics Curriculum Literature Database, Curriculum Research Instrument Database, K-12 Mathematics Textbook Database*, and the *State Mathematics Standards Database*. The first two of these comprised some of the early work of the CSMC as faculty and students worked to establish databases that were searchable, useful, and discretionary with respect to articles focused, in a broad sense, on K-12 mathematics curriculum. Thus, the *Literature Database* (the larger of the two) consists of abstracts of articles that have appeared in past education research journals, but does not include unpublished dissertation work. CSMC faculty and students also elected not to include standardized testing instruments, as information related to those are easily searchable on their own websites and were not likely to be tied to research focused on curriculum. Since its development, the *Literature Database* has been maintained by graduate students from the three CSMC home institutions[3] who monitor various journals for articles related to curriculum for potential inclusion. They either upload an existing abstract or write one if none exists. Articles must have a focus on mathematics curriculum research, and may range across at least one of six broad areas defined by the CSMC research agenda: (1) curriculum design and analysis, (2) curriculum adoption and enactment, (3) curriculum and student learning, (4) curriculum and teacher knowledge and beliefs assessment, (5) history and status of curriculum, and (6) policy and curriculum. As a companion to the *Literature Database*, a second database, titled the *Curriculum Research Instrument Database*, was developed from a search of relevant literature and independent reviews of articles. This instrument database is the focus of the remainder of this chapter.

STRUCTURE OF
THE CURRICULUM RESEARCH INSTRUMENT DATABASE

The *Instrument Database*[4] consists of entries describing tools that have been used to conduct curriculum research. However, not all of the included tools focus on curriculum implementation (i.e., the enacted curriculum). In the following descriptions of the structure and content of the database, we speak, in general, about the identified tools, then focus more carefully on those directed at research involving the enacted curriculum.

During the development of the *Instrument Database*, CSMC researchers collected many of the instruments from websites, articles, National Science Foundation (NSF) reports, dissertations,[5] or directly from the researchers responsible for their development. However, the database itself is neither a repository nor a clearinghouse for these tools. CSMC researchers felt it was important for authors to retain control of their own instruments; so, interested researchers are expected to initiate direct contact with the instrument developers in order to obtain the tools for use or modification.

The following journals are the most represented in the *Instrument Database*: *Cognition and Instruction, Curriculum Inquiry, Educational Researcher, Educational Studies in Mathematics, International Journal of Educational Research, Journal of Educational Research, Journal of Mathematics Education Leadership, Journal of Mathematics Teacher Education, Journal for Research in Mathematics Education,* and *School Science and Mathematics*. While analyzing the research pieces in the *Literature Database*, we also followed up on references to other articles that either used or discussed curriculum-related instrumentation. As a result of these efforts, the *Instrument Database* houses only references to research published in English, with primary emphasis on work concerning curriculum research in the United States.

Each entry in the database consists of an *Instrument Evaluation Summary* (IES) that contains the following information summarized from a published article:

- title and type of instrument,
- the author(s) and institution where the instrument was developed,
- the intended research subjects or participants whom the instrument was designed to study,
- a brief description of the instrument, including where it has been cited in the literature,
- constructs the instrument purports to measure,
- the time and cost it takes to administer the instrument,
- training (if any) required for administration, and
- any information reported regarding validity and reliability of the instrument.

A sample of an IES for a tool developed by Reys (2006) for the *Middle School Mathematics Study* (MS)^2 is shown in Figure 5.1. This three-year study examined the impact of middle school mathematics curricula, developed in response to the National Council of Teachers of Mathematics' (1989, 2000) *Standards* documents on student achievement and the classroom-learning environment. Several instruments were developed and used in conjunction to address the project's research questions. The identified instrument is a Textbook-Use Diary that was used to collect data from inservice teachers.

The *Instrument Database* is searchable by key word, type of instrument, and research focus. A typical screen layout of how one would search for interview protocols used with elementary students is given in Figure 5.2. This particular search yielded three references shown at the bottom of the Figure under *Title of Instrument*. An additional click on a select instrument title will take the user to the IES information for that tool. The user can also type in search words (e.g., research keywords or an author's last name) and run the query with or without *Type of Instrument* or *Research Subject* specified. Researchers using this database may start with this search feature as they look for instruments with particular constraints that match their interests.

Figure 5.1. Example of an Instrument Evaluation Summary in the *Instrument Database*

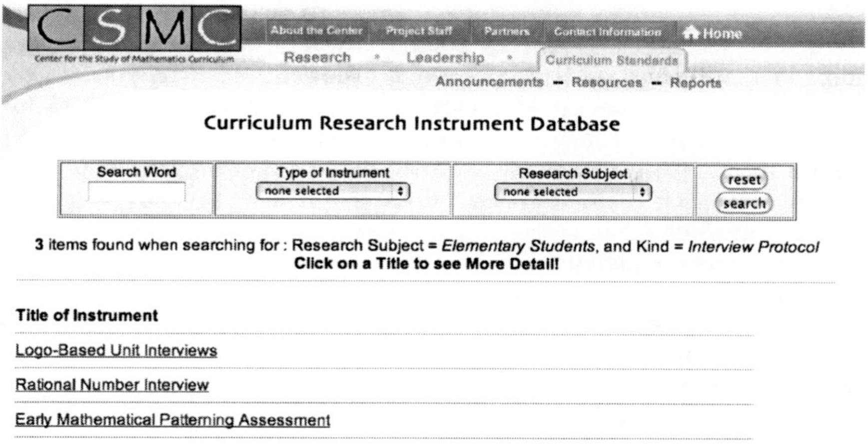

Figure 5.2. An example of a search within the *Instrument Database*.

Content of the Database

The *Instrument Database* consists of entries that describe tools used to conduct curriculum research. Historically, highlighting the *tools* piece of school mathematics curriculum research has been somewhat ancillary to more focused attention on important questions related to curriculum change, implementation or enactment, or student outcomes. Fullan and Pomfret (1977) examined curriculum change in the 1970s and reported the use of observations, questionnaires, and document analysis as typical ways in which data were collected. The National Research Council's (NRC, 2004) evaluation of curriculum evaluations noted the wide range of methodologies and instruments used in those studies to determine curricular effectiveness. Some reported studies relied on a single instrument while others used a variety of measures to determine outcomes. However, when the NRC summarized the evaluations based on *implementation fidelity* (i.e., enactment), more than half (53%) "recorded no information on the issue" (p. 115). Thus, reports of tools used in studying this aspect of curriculum research seemed to be lacking in many cases.

Two more recent reviews have demonstrated that some progress has been made in developing and using new tools to study curriculum enactment. Remillard (2005), in her review of mathematics curriculum research three decades after Fullan and Pomfret, described the use of tools similar to those reported as common among researchers in the 1970s, but also described uses of student and teacher interviews and classroom video analyses as important contributions to the instrument repertoire. Most recently, O'Donnell's (2008) study of links between curriculum implementation and student achievement showed similar types of instruments standard among curriculum researchers. A companion book to this volume, *Approaches to Studying the Enacted Curriculum*, by the CSMC Tools Group (Heck, Chval, Weiss, & Ziebarth, 2012) more fully explores these earlier studies; it features independent chapters by several prominent curriculum researchers or teams of researchers who discuss the development of their instruments used in more recent curriculum work. Of note in a number of their chapters are the *combinations* of instruments needed to understand the complex phenomena involved in answering questions about mathematics curriculum research, particularly those focused on implementation issues (NRC, 2004).

The *Instrument Database* allows us to begin to examine tools broadly used in curriculum research, while identifying those instruments that were developed to focus specifically on questions of curriculum enactment. In our search for curriculum research tools, five broad categories emerged that warrant some brief discussion. In general, a pattern similar to those of previous researchers appeared, with the dominant tools of choice still in the categories of surveys, observation instruments, and interview protocols. However, curriculum research and associated instruments over the past two decades have taken some new directions, mainly with a focus on comparing curricula through textbook analyses and making links to student outcomes (e.g., assessment and accountability). Table 5.1 summarizes the major categories of tools in the *Instrument Database* at the time of this analysis, separated into five major groupings. For ease in comparing these types of tools, the instruments are further disaggregated by research focus into categories of students or educators depending on the targeted population for the data collection.

In the sections that follow, we elaborate on each of the categories of tools introduced in Table 5.1. First, we highlight those that focus most closely on the enacted curriculum—surveys, observation protocols, and interview protocols—and describe several examples of each. Student assessment instruments and curriculum analysis tools are discussed last. Although these latter instruments are less likely to target the enacted curriculum, they can be effectively used in concert with other instruments that do target the enacted curriculum.

Surveys

As shown in Table 5.1, surveys are by far the instrument of choice in curriculum research. Surveys in their variety of forms are broadly useful

Table 5.1. Number of Instruments by Type and Educational Focus[1]

Research Participants	Type of Instrument					
	Survey	Observation Protocol	Interview Protocol	Student Assessment	Other	Total
Students						
All Grades	0	0	0	3	0	3
Elementary	0	0	3	8	0	11
Middle School	4	1	2	1	1	9
High School	8	1	2	7	2	20
Educators						
Preservice Teachers	2	0	0	n/a	1	3
Inservice Teachers	38	20	11	n/a	3	72
Administrators	4	0	0	n/a	0	4
Total	56	22	18	19	17*	122*

1. We conducted our analysis during the fall of 2011 and the spring of 2012.
*There are 10 additional instruments within the database (as a type of instrument) for curriculum analysis that are not directed at a particular population.

for data collection across a range of target audiences. They are fairly easy to develop, can attend to a range of constructs, are generally cost effective, and can be administered at a distance. Among the drawbacks can be the limitation of possible responses to items, misinterpretation of questions and responses, user differences in scale interpretation, nonresponse, and accuracy or validation of classroom data collection. This last issue seems particularly relevant to research efforts directed towards the enacted curriculum. To what extent can surveys *alone* accurately capture what teachers do with curriculum materials while teaching? As we will see below, most surveys form part of a broader package of instruments used in conjunction to capture classroom teaching activities and behaviors.

For the purpose of combining instruments in the *Instrument Database* into searchable categories, we grouped various types of surveys together, regardless of how researchers designated them in terms of title. Thus, the Survey category broadly includes thoughts, practices, opinions, and other information (e.g., demographics) that respondents or research subjects can record to help answer research questions. Many are called "surveys" (e.g., Initial Teacher Survey, Math Opinion Survey, Wasman Teacher Survey), but others are titled "questionnaires" (e.g., LSC Teacher Questionnaire, MIC School Context Questionnaire), "indexes," "scales," "profiles," "records," or "ratings." Although most of the surveys are intended for completion in a single administration (as with pre- and/or post-surveys), we also included longer-term or continuous data collection instruments that are identified as teacher or teaching logs, planning logs, coverage logs, and textbook-use diaries. In many ways, the log and diary instruments serve the same purpose for collecting curriculum use data routinely, on a daily or weekly timeframe, as do the one-time survey instruments administered at the end of a semester or year. The Opportunity-to-Learn Checklist (cf. Thompson & Senk, 2001) or one of the various topic lists from a version of the Surveys of Enacted Curriculum developed by Porter, Schmidt, Floden, and Freeman (1978)[6] are familiar examples of single administration tools for recording classroom content coverage. Table 5.2 presents a summary of common variables in the surveys referenced in the *Instrument Database* at the time of this analysis.

In our review of the *Instrument Database*, we identified 18 of the 56 (32%) survey-type tools as direct attempts to measure the enacted curriculum. In addition, 36 (64%) of the instruments are part of a larger package of tools identified with at least 10 different research projects or research groups. Some are focused on a particular curriculum and were developed as part of the evaluations connected with NSF-funded or other curriculum development projects. For example, the Opportunity-to-Learn Checklist (Thompson & Senk, 2001) previously noted is one of five survey-type tools developed for use during evaluation work on the textbooks developed by the University of Chicago School Mathematics Project (UCSMP). This tool, in combination with a UCSMP Chapter Evaluation Form, collects data on content coverage by topic, lesson, and unit to enable the researchers to gauge what subset of the materials was *actually* taught to or reviewed with students in the classroom. Pre- and post-opinion surveys

Table 5.2. **Common Variables Measured by Surveys**

Target Population	Inservice teachers
	Students, as perceived by the teachers
Constructs Measured	Classroom culture
	Classroom discourse: teacher-student, student-student
	Implementation of a particular curricular program/intervention
	Mathematical tasks, cognitive demand
	Student actions and engagement in tasks/problem solving
	Teachers' instructional decisions
	Teachers' pedagogy
Time/Duration	Single administration
	Long-term, continuous/recurring

and a Teacher Questionnaire round out this suite of survey tools for each textbook evaluated.

The *Instrument Database* also contains several different surveys identified with evaluation work on two NSF-funded curriculum development projects, *Core-Plus Mathematics* (two focused on enactment) and *Connected Mathematics* (three focused on enactment from three different research groups). The largest bank of survey instruments ($n = 10$) is identified with a cross-sectional/longitudinal study of the *Mathematics in Context* middle school curriculum. Developed by the Wisconsin Center for Educational Research (WCER), the teacher log instrument had teachers report on enactment data such as content taught, forms of instruction, room arrangement, student activities, and assessments used (Shafer, Wagner, & Davis, 1997). Figure 5.3 presents some items from these surveys that target student class work in collaborative settings. Supporting surveys for this instrument include school and district profiles and several types of teacher questionnaires. Separate surveys by grade level (K-8 and 6-12) multiply the number of instruments used in this project.

AMOUNT OF INSTRUCTIONAL TIME *(in pairs or small groups)*
0 - **None**
1 - **Little** *(10% or less of instructional time in pairs or small groups)*
2 - **Some** *(11-25 % of instructional time in pairs or small groups)*
3 - **Moderate** *(26-50% of instructional time in pairs or small groups)*
4 - **Considerable** *(51-75% of instructional time in pairs or small groups)*
5 - **Almost all** *(more than 75% of instructional time in pairs or small groups)*

When students in the target class work *in pairs or small groups* on math exercises, problems, investigations, or tasks while in the classroom, how much time do they:

	None	Little	Some	Moderate	Considerable	Almost all
45 Solve *word problems* from a textbook or worksheet.	⓪	①	②	③	④	⑤
46 Solve non-routine mathematical problems (for example, problems that require novel or non-formulaic thinking).	⓪	①	②	③	④	⑤
47 Talk about their reasoning or thinking in solving a problem.	⓪	①	②	③	④	⑤
48 Apply mathematical concepts to "real-world" problems.	⓪	①	②	③	④	⑤
49 Make estimates, predictions or hypotheses.	⓪	①	②	③	④	⑤
50 Analyze data to make inferences or draw conclusions.	⓪	①	②	③	④	⑤
51 Work on a problem that takes at least 45 minutes to solve.	⓪	①	②	③	④	⑤
52 Complete or conduct proofs or demonstrations of their mathematical reasoning.	⓪	①	②	③	④	⑤

Figure 5.3. An excerpt from one of the WCER surveys.

As another example, four different survey instruments are included from Reys' (2005) Middle School Mathematics Study (MS)^2 project, including a *Table of Contents Record* and a *Textbook Use Diary* which, combined with other surveys and observation tools, were used to examine the enactment of several middle school curricula across multiple national sites. For example, the *Textbook Use Diary* asked teachers to describe their textbook use on a daily basis for three 10-day periods throughout the school year.

We close this brief overview of the survey section of the *Instrument Database* by reemphasizing that a main goal of this project has been to provide a vehicle for sharing and highlighting the curriculum enactment research of others and the tools they have developed. Our survey category emphasizes this theme of sharing instruments in another way, adapting instruments from one project for use in another. For example, we note that ten of the survey instruments in the database were developed by Horizon Research, Inc., specifically for two of its high-profile projects: the 2000 National Survey of Science and Mathematics Education; and the evaluation of the NSF Local Systemic Change (LSC) Initiative that was conducted over several years in the late 1990s. Although we did not identify these ten surveys as directly related to curriculum enactment, they are part of a larger package of instruments, including observation protocols that collect data on enactment at the classroom level (albeit more focused on pedagogical changes). Because of the national scale of the Horizon, Inc. projects, the instruments have been readily used or adapted for other evaluations and studies.

Other examples of adapting existing instruments for new research can be seen in Reys' adaptation of the WCER teachers' logs for the (MS)^2 study. In turn, some of the WCER instruments were influenced by Norman Webb's evaluation work with the NSF-funded *Interactive Mathematics Program* (IMP) (J. D. Davis, personal communication, February 8, 2012). Additionally, Schoen's (cf. Schoen, Cebulla, Finn, & Fi, 2003) Core-Plus Mathematics Project survey and observation tools built on the work of the 1990s *Quasar* Project (cf. Silver, 1994).

Finally, several instrument summaries in the database acknowledge the influence of the Concerns-Based Adoption Model (CBAM) developed in the 1970s by Gene Hall and his colleagues (cf. Hord, Stiegelbauer, Hall, & George, 2006). Huntley's (2009) work on developing Innovation Configuration Maps to study middle school curriculum enactment associated with *Connected Mathematics* and *Math Thematics* is perhaps the most recent example. The Concerns Based Adoption Model (CBAM) developed instruments for studying the process of implementing educational change by teachers and by persons acting in change-facilitating roles.

Observation Protocols

By their designed intent, all 22 observation protocols included in the database at the time of this analysis were related to documenting the enacted curriculum—what takes place in the classroom. Each of these instruments addressed the key players of the enacted curriculum—teachers, students, and tasks (e.g., curriculum materials)—with varied degrees of emphasis and focus. As a collection, these instruments spanned several key topics related to the enacted curriculum, including implementation, interactions (e.g., teacher-student, student-student, student-textbook), instruction, discourse, time, tasks/textbooks, cognitive demand, content, reform, and culture/environment/context of the classroom. Each of these topics can be used in the search field on the *Instrument Database* to further refine the set of Observation Protocols. Table 5.3 summarizes some key constructs that are measured across this collection of observation protocols.

Many protocols emphasize teacher-student interaction and student-student interaction, including classroom discourse, roles, and behaviors (e.g., Matsumura et al., 2006). Others focus more prominently on the teacher's implementation of mathematics content and pedagogy. For example, Fuchs, Fuchs, Hamlett, Phillips, and Bentz (1994) focus on fidelity of use of curriculum materials. Also, consider the sample shown in Figure 5.4 from the Instructional Quality Assessment (IQA) lesson observations rubric (Matsumura et al., 2006). According to M. D. Boston (personal communication, June 2, 2011) "The IQA assesses the level of cognitive demand of instructional tasks at different points throughout an instructional episode, the rigor of classroom talk, and the rigor of teachers'

Table 5.3. Common Variables Measured by Observation Protocols

Key Players	Teachers
	Students
	Tasks (curriculum materials)
Type of Interaction	Teacher-student
	Student-student
	Student-textbook
Instruction	Pedagogy (e.g., student-centered)
	Time
	Discourse and questioning
	Tasks and textbooks
	Roles and behaviors
	Classroom culture/environment
Mathematics	Content
	Task Selection
	Cognitive Demand

expectations." Finally, a small subset of the protocols puts students as the main focus with attention to student engagement with tasks. For example, Boaler's (1998) instrument measures variables of cognitive demand, motivation, and focus. It is significant to note that for 20 of the 22 instruments, the primary research subject was the inservice teacher, with the only other 2 instruments concerned with the secondary student.

There is great variety among this collection of instruments in how the authors attend to the links between the enacted curriculum and other phases of curriculum use: written, intended, and student learning (cf. Stein, Remillard, & Smith, 2007). Some protocols are designed for classroom observations in which particular curricula and textbooks are used (e.g., Core-Plus Mathematics Project, 2003), while others remain more curriculum-neutral (e.g., Artzt & Armour-Thomas, 1999). With respect to the written curriculum, some protocols emphasize the use of textbooks and the ways in which teachers and students interact with these materials (e.g., Reys, 2005; Tarr, Reys, Reys, Chávez, Shih, & Osterlind, 2008). Finally, the theme of "reform" permeates some observation protocols, such as those focused on reform-oriented curricula (e.g., Michlin, Seppanen, & Sheldon, 2001) or reform-oriented teaching practices (e.g., Schoen, Cebulla, Finn, & Fi, 2003; and the LSC protocols). The term "reform" in these contexts refers to curriculum and teaching

Rubric AR-Q: Questioning	
4	The teacher **consistently** asks academically relevant questions that provide opportunities for students to elaborate and explain their mathematical work and thinking (probing, generating discussion), identify and describe the important mathematical ideas in the lesson, or make connections between ideas, representations, or strategies (exploring mathematical meanings and relationships).
3	At least 2 times during the lesson, the teacher asks academically relevant questions (probing, generating discussion, exploring mathematical meanings and relationships).
2	There are one or more superficial, trivial, or formulaic efforts to ask academically relevant questions probing, generating discussion, exploring mathematical meanings and relationships) (i.e., every student is asked the same question or set of questions) or to ask students to explain their reasoning; OR only one (1) strong effort is made to ask academically relevant questions.
1	The teacher asks procedural or factual questions that elicit mathematical facts or procedure or require brief, single word responses.
0	The teacher did not ask questions during the lesson, or the teacher's questions were not relevant to the mathematics in the lesson.
N/A	Reason:

Figure 5.4. An excerpt from the IQA lesson observation rater packet showing a rubric for teacher questioning

practices that reflect various interpretations of the NCTM *Standards* (1989, 2000) documents.[7]

Despite these commonalities in studying reform or reform-oriented practices, the ways in which these constructs are measured varies. For instance, Schoen's Observation Protocol (Schoen et al., 2003) involves giving a fidelity rating for the extent to which a teacher exhibits reform-teaching practices. The criteria for reform-teaching practices were derived from the literature and aligned with the intended curriculum and NCTM's *Standards*. These practices include open-ended questioning, student as math authority, student learning from investigations, class organization, collaborative work among students, tools used appropriately, and mathematical understanding of the big idea in the respective investigation. To use this protocol, an observer records comments on these predetermined categories related to reform-oriented teaching, then retrospectively uses their general impression of the teaching episode to inform their rating of a teacher's practice and classroom culture on a holistic level (excellent, good, fair, poor). Schoen's protocol was one of several observation tools that specifically focused on the enactment of reform curricula. It should be noted that each study that referenced such an instrument mentioned the need for a significant amount of training on the part of the user.

For a different example, Boaler's Observation Protocol (Boaler, 1998) focuses on the role of students in the classroom and their engagement with mathematics. As part of this protocol, a researcher assumes a participant-observer role in the classroom and documents student time-on-task, student approaches to mathematics, and student behaviors (e.g., rule-following behavior). The observer also documents student activities at the beginning, middle, and end of the instructional session to capture experiences of the students. The role of the observer in the classroom setting while instruction is occurring is particularly significant for this protocol. That is, it would not be feasible to retrospectively analyze curricular enactment using this instrument.

One obvious point of contrast between these two protocols is that Schoen's Protocol takes a more teacher-centered approach with specific criteria in mind to capture both instructional materials and pedagogy, while Boaler's protocol focuses on students and their actions during the enacted curriculum. Both protocols require differing degrees of interpretation on the part of the observer, with Schoen's involving a type of "quality" rating based on specific criteria, and Boaler's involving documentation of how students are interacting with the mathematical materials. The nature of this interpretation is closely linked to one's background and experience (or training) in using such tools. Moreover, both observation instruments are part of suites of tools for each respective study, exemplifying the interconnected nature and complexity of studying curricular enactment.

In summary, the Observation Protocols housed on the CSMC *Instrument Database* represent a diverse collection of tools that can be used to document and study the enacted curriculum by measuring a variety of variables. Tools differ in the constructs they measure, their relationship to other phases of curriculum use, the open-ended or close-ended nature of the protocol, and their emphasis on various players in the enacted curriculum—teachers, students, and tasks. This discussion has highlighted some more specific ways that this collection of instruments can be characterized, which can inform searches in the database, and intent of use.

Interview Protocols

Eighteen of the instruments in the database fall under the category of interview protocols. These instruments focus on a broad spectrum of constructs, including teacher thoughts regarding lesson planning; their beliefs about mathematics, pedagogy, and curriculum; teacher reflections on their own practice; and student perspectives about mathematics and their own learning processes. At the time of this analysis, there were no interview protocols referenced in the database that attend to the perspectives of preservice teachers[8] or school administrators. As noted with other types of instruments, some of the interview protocols belong to suites of tools aligned with a particular curriculum project (Collopy, 2003; Huntley, Rasmussen, Villarubi, Sangtong, & Fey, 2000) or particular interventions related to curriculum (Fuchs et al., 1994; Papic, Mulligan, & Mitchelmore, 2011). Only two of the interview protocols fall short of clear relation to the enacted curriculum. Table 5.4 presents the most common items examined in the interview protocols referenced in the *Instrument Database*.

Table 5.4. Common Items Measured by Interview Protocols

Target Population	Inservice teachers
Constructs Measured	Teacher mathematical knowledge
	Teacher pedagogical knowledge
	Beliefs about teaching mathematics
	Beliefs about student motivation
	Goals for instruction
	Pacing of content and curriculum coverage*
	Use of mathematical technologies during instruction*
	Implementation of NCTM *Standards*-based curricula*

* These items speak directly to elements in the enacted mathematics curriculum

Among the 11 protocols that target inservice teachers, those most strongly linked to the enacted curriculum prompt the instructors to reflect on their teaching. Typically, the protocols prompt teachers to think about a particular teaching occurrence and explain their process of instruction, how the instruction aligned with the textbook and/or lesson plans, or things they might adapt in future instruction. These protocols often pair with classroom observations (sometimes both pre- and post-observation interviews), so that researchers can confirm and probe into patterns that emerge during the observations (e.g., Weiss, Pasley, Smith, Banilower, & Heck, 2003). Some researchers also collect video data for teachers to view and report their reactions and reflections in interviews afterwards (e.g., Artzt & Armour-Thomas, 1999; Middleton, 1995), allowing for triangulation among data sources. The duration of these interviews ranges from a few minutes (in the case of pre-observation interviews) to as long as an hour (in the case of final or more in-depth interviews). Figure 5.5 presents an example of questions asked during teacher interviews.

Many of the interview instruments probe teachers' broad knowledge and beliefs about content and curriculum, pacing, student motivation, and the use of mathematical technologies. Doing so helps researchers see beyond the scope that is captured in the observation protocols and allows for a wider measure of consistency between teachers' goals and beliefs,

---

B. **Content/Topic**

7. What led you to teach the mathematics/science topics/concepts/skills in this lesson?

   (Use the following probes, as needed, so you can assess the extent of importance of each of these influences:)

   Is it included in the state/district curriculum/course of study?
   If yes, or previously implied: How important was that in your decision to teach this topic?

   Is it included in a state/district mathematics/science assessment? What are the consequences if students don't do well on the test?
   If yes, or previously implied: How important were these tests in your decision to teach this topic?

   Is it included in an assigned textbook or program designated for this class?
   If yes, or previously implied: How important was that in your decision to teach this topic?

Source: Weiss, Pasley, Smith, Banilower, and Heck (2003).

Figure 5.5. Example of questions from a teacher interview protocol.

their actual instruction, and the goals of a particular curricular program. For example, researchers can investigate teachers' beliefs about the use of mathematical technologies under curricular programs or interventions that utilize such technologies for particular mathematical purposes (see Clements, Battista, Sarama, Swaminathan, & McMillen, 1997; Huntley et al., 2000).

At the time of this analysis, students were the targeted research participants in seven of the interview protocols, and there was at least one instrument identified in the database for each of the elementary, middle school, and high school grade bands. In these student interview protocols, students were most often prompted to share their perspectives regarding their learning under particular curriculum projects or instructional interventions (e.g., Clements et al., 1997; Papic et al., 2011). Typically, researchers designed student interviews to last no more than 30 minutes, a compromise between depth of data collection and the constraint of time.

Artzt and Armour-Thomas (1999) use teacher-focused interview protocols in a series of three interviews, each occurring after a period of instruction. The first is a structured protocol that measures teachers' first reflections after an episode of instruction, and should be used as soon after instruction as possible. During an opportunity for teachers to watch video of themselves teaching, the second protocol allows teachers to identify and comment on those moments during instruction when they made specific decisions in adjusting their instruction while teaching; this process is referred to as *stimulated-recall*. The final instrument of the set targets teachers' responses to their own observed teaching and allows them to reflect on items they might not have considered had they not seen the videos or that they would change in future instruction.

The *Rational Number Test* of Moss and Case (1999) represents a task-based interview protocol that attends to student assessment at the same time. Here, the researchers would also implement pre- and post-tests surrounding the instructional intervention, focusing on rational numbers. This particular instrument is unique in its specific content focus on rational numbers, which helps to illuminate the diversity in the scope of the *Instrument Database*. Its relation to the enacted curriculum lies in its design to assess student learning outcomes as the result of instruction on rational numbers. Researchers would use this protocol in examining the learning achieved by students (possibly in control and treatment groups) or to compare student work at the task/problem level. It is noteworthy that the Rational Number Test is one of only three interview protocols referenced in the database for use with elementary-level students.

The two protocols referenced in the preceding paragraphs differ in several respects. First, they target different research subjects (teachers versus elementary students). Second, they align with different research designs (investigative and descriptive versus evaluative and experimental). Further, Artzt's and Armour-Thomas' protocol attends to pedagogy, while Moss' and Case's protocol attends to the specific mathematical content of rational numbers. With these differences noted, it is worth stating that these protocols could be adapted and used in a single research project that combines the teacher's reflections with the actual learning of the students with respect to the content of rational numbers.

To summarize, the interview protocols referenced in the CSMC *Instrument Database* represent a broad collection of instruments that researchers may utilize in studying different aspects of the enacted curriculum and parallel the scope of the surveys and observational protocols in the database. Researchers looking to examine the impact of a particular textbook or intervention may find adaptable materials referenced in the database, which can serve as a starting point for future tool development.

Curriculum Analysis Tools and Student Assessment Instruments

Examining instructional materials and student assessments, although not directly linked to the enacted curriculum, can help to promote robust descriptions of factors that influence teachers' enactments. For example, a curriculum analysis might pair well with teacher reflections that refer directly to elements of the instructional materials that receive (or do not receive) attention during interviews. As a second example, researchers might follow-up on a classroom observation sometime later with a student assessment in order to investigate the durability of specific learning outcomes that seemed to emerge in the observed lessons.

Curriculum comparisons have become more common since the emergence of several new instructional materials representing authors' different interpretations of the National Council of Teachers of Mathematics' *Standards* (NCTM, 1989, 2000) documents and the rising popularity of some non-U.S. textbooks (e.g., Singapore Math). The development of various curriculum analysis tools that seek to compare these alternatives to commercially-developed texts has grown. There are enough of these tools identified to warrant a separate category within the database and we have identified 10 that differ in focus and approach. One example illustrating the focus on non-U.S. curriculum comparisons is the Middle School Mathematics Comparisons tool developed by Adams and colleagues (2000) at the University of Washington. This instrument ranked three middle school curricula (*Connected Mathematics, Mathematics in Con-*

text, and Singapore mathematics) on a scale from 0 to 3 on their alignment with the NCTM's (2000) revised content and process standards.

In addition to the nationally recognized curriculum analysis work of Project 2061 (cf. American Association for the Advancement of Science [AAAS], 2004), less well-known curriculum analysis instruments have focused their comparative analyses on other constructs, such as teacher behaviors suggested in elementary textbooks (Black, 1986), special education (Fischer, 1997), questioning levels based on Bloom's Taxonomy (Jacobbe, 2009), and values portrayals (Seah & Bishop, 2000). Although these tools do not directly examine enactment of materials in the classroom, they help researchers understand the different content and pedagogical demands and expectations that various curricula emphasize and may serve as a valuable precursor to the use of other tools for data collection.

In the category of student assessment instruments, CSMC researchers have consciously avoided many of the high-profile testing instruments, although the database does include a number of assessments authored by such companies as ACT (e.g., Work Keys, PACT+). As noted previously, information on several of the well-known testing instruments is already easily accessible, often directly on the governing institutions' websites. The CSMC *Instrument Database* has sought to provide information on those instruments that are more heavily focused on curriculum and those that were developed unique to specific projects and that might not be widely distributed.

Comparative studies cited in the NRC (2004) report showed that many curriculum evaluations employed standardized tests as a popular tool to measure effectiveness, yet lacking in these comparisons was data on curriculum implementation fidelity. O'Donnell's (2008) more recent review has made progress at connecting fidelity and outcome measure variables and several recent projects that use multiple instruments to measure enactment report working on attempts to connect/combine them in ways that will make various outcome measures (i.e., student achievement) meaningful (cf. Heck et al., 2012). In summarizing this category within the *Instrument Database*, we note that three of the high school assessment instruments were developed by the UCSMP evaluators (cf. Thompson & Senk, 2001) to assess several algebra-related constructs; one is Huntley and colleagues' (2000) assessment of algebraic thinking with the Core-Plus Mathematics Project and comparison students, and another is Tarr and colleagues' (2008) use of the Balanced Assessment in Mathematics instrument in the $(MS)^2$ study. Some of the assessment instruments unique to elementary students focus on measuring their understanding of multiplication (cf. Smith & Smith, 2006), using building blocks to measure thinking and learning along developmental progressions (cf. Clements & Sarama, 2007), and exploring geometric concepts of length and distance using Geo-Logo (cf. Clements et al., 1997) to measure curriculum effects. However, our investigations into those studies focused at the elementary level do not clearly reveal the extent to which fidelity of implementation was a key variable of interest in determining the student outcomes.

GAPS AND NEXT STEPS

The *Instrument Database* offers an expansive collection of research tools, but is not exhaustive. As we have outlined, surveys, observation protocols, and interviews remain, alone or in conjunction, the standard tools of choice in research on the enacted curriculum. The *Instrument Database* has been a first attempt at surveying the landscape of tools developed and used during the past few decades, but it continues to be a work in progress. Certainly, gaps exist with respect to the instruments included in the database. One area of improvement of the database would be to extend beyond the current collection of tools to include more recently published instruments, but also identifying the work of researchers who report results of projects with only hints at the tools they used to collect their data. Open access to an instrument database is another viable way to encourage collaboration and testing of instruments in a variety of research settings.[9]

Moreover, should more venues be found for reporting particular protocols or instruments, the purpose and procedures for using those instruments would also be helpful. While some cursory information on constructs measured and required training are included in this database, evidence of instrument validity and reliability is woefully lacking in the referenced literature. This creates a second gap in the *Instrument Database*, as we have had insufficient information to include in the IES summaries. Some research studies specifically address the inter-rater reliability efforts related to testing their instruments; many studies do not include such efforts. To improve the validity of the instruments used in studies, such testing should be included in all research reports.

In addition to the perceived lack of specific attention to validity and reliability measures, other gaps exist with respect to the literature base. Some studies fall short of clarifying the purpose and nature of the instrumentation that was used. Such information is necessary for other investigators who seek to make informed decisions about using and/or adapting existing instruments. In addition, the guiding framework or lens of the tool's author(s), along with the research questions being investigated, could provide important information for a user to judge the purpose and intent of the tool. Finally, as noted in numerous examples, increasingly we see new curriculum implementation research that requires multiple instruments to deal with the complex phenomena of what teachers do with curriculum materials in the classroom. This may include protocols that, on their own, do not target the enacted curriculum, yet become useful when used in concert with instruments that do. We believe it is important for this trend of using multiple tools to continue. Combining information in usable ways that can inform practice remains a challenge both within the database and in the field. With the diversity and variety of available curriculum materials, more curriculum-neutral protocols may be desirable for ease in use and flexibility across studies.

We close by identifying some next steps that are related to the CSMC Research Framework.[10] From our analysis of the *Instrument Database*, it seems as though most of the surveys, observation protocols, and interview protocols that have been compiled are directed at the audience of the inservice teacher and are concerned with measuring variables associated with the implemented curriculum (shown in the model as the *Teacher: Implemented Curriculum* node). Also, as alluded to previously, often a collection of instruments is needed to fully understand the complexities of the implemented curriculum. In light of this, gaps seem to be most prominent in understanding how this type of curriculum research is linked to textbook analysis (the *Textbook Curriculum* node in the framework) and also the learned curriculum (the *Student Learned Curriculum* node in the framework).[11] By directing research efforts toward such links (the arrows in the Research Framework) between the phases of curriculum, the utility of curriculum instruments might be better realized.

NOTES

1. The CSMC was funded in 2004 by the National Science Foundation under Grant #ESI-0333879 as one of 15 Centers for Learning and Teaching (CLTs), with the focus of CSMC on K-12 mathematics curriculum. More information about the Center can be obtained at http://www.mathcurriculumcenter.org/about_the_center.php.
2. A significant part of the early work on the interfaces of the databases was conducted by CSMC graduate fellows Dana Cox and Óscar Chávez.
3. Western Michigan University, Michigan State University, and University of Missouri, Columbia are the three institutions that provided ongoing support for graduate students. The University of Chicago and Horizon Research, Inc. have also been research partner institutions throughout the duration of the grant.
4. This database is accessible at http://www.mathcurriculumcenter.org/InstrumentDatabaseNew.php.

5. Dissertations were included in searches for tools in the *Instrument Database*, but not in the CSMC *Literature Database*.
6. A fuller description of both of these tools is given in Heck et al. (2012) and both are entries in the database.
7. The authors of this chapter fully recognize the problematic nature of terms, such as "reform," that vary among curriculum writers and publishers who argue for alignment of their work with either of the NCTM curriculum publications in 1989 or 2000. In many recent cases, the term has been replaced by "standards-based," which can have connotations well beyond those articulated in the NCTM documents. In describing their instruments, we have tried to retain the authors' intent as conveyed in the instrument title or context of the research conducted.
8. This identifies an example of gaps in the research that may warrant investigation.
9. Authors are encouraged to share references to other curriculum instruments with steven.ziebarth@wmich.edu.
10. Chapter 6 of this volume, by Remillard and Heck, elaborates on this framework.
11. See Center for the Study of Mathematics Curriculum (2013a, 2013b).

REFERENCES

American Association for the Advancement of Science (AAAS). (2004). *The Project 2061 analysis procedure for mathematics curriculum materials*. Retrieved from http://www.project2061.org/publications/textbook/algebra/report/analysis.htm

Adams, L., Tung, K. K., Warfield, V. M., Knaub, K., Mudavanhu, B., & Yong, D. (2000). Middle school mathematics comparisons for Singapore mathematics, *Connected Mathematics Program*, and *Mathematics in Context*. Report to the National Science Foundation. (Unpublished manuscript)

Artzt, A. F., & Armour-Thomas, E. (1999). A cognitive model for examining teachers' instructional practices in mathematics: A guide for facilitating teacher reflection. *Educational Studies in Mathematics, 40*, 211-235.

Black, M. L. (1986). *Content analysis of five elementary textbook series* (Unpublished dissertation). University of Illinois at Urbana-Champaign.

Boaler, J. (1998). Open and closed mathematics: Student experiences and understandings. *Journal for Research in Mathematics Education, 29*(1), 41-62.

Center for the Study of Mathematics Curriculum. (2013a). CSMC: About the Center. Retrieved from http://www.mathcurriculumcenter.org/about_the_center.php

Center for the Study of Mathematics Curriculum. (2013b). CSMC: Curriculum Databases. Retrieved from http://www.mathcurriculumcenter.org/resources_databases.php

Clements, D. H., Battista, M. T., Sarama, J., Swaminathan, S., & McMillen, S. (1997). Students' development of length concepts in a Logo-based unit on geometric paths. *Journal for Research in Mathematics Education, 28*(1), 70-95.

Clements, D. H., & Sarama, J. (2007). Effects of a preschool mathematics curriculum: Summative research on the Building Blocks Project. *Journal for Research in Mathematics Education, 38*(2), 136-163.

Collopy, R. (2003). Curriculum materials as a professional development tool: How a mathematics textbook affected two teachers' learning. *Elementary School Journal, 103*(3), 287-311.

Core-Plus Mathematics Project. (2003). *CPMP class observation: Revision study format.* CPMP: Western Michigan University.

Fischer, T. A. (1997). *A content analysis of United States math textbooks, 1966-1996 from a special education perspective* (Unpublished dissertation). University of Wisconsin-Madison.

Fuchs, L. S., Fuchs, D., Hamlett, C. L., Phillips, N. B., & Bentz, J. (1994). Class-wide curriculum-based measurement: Helping general educators meet the challenge of student diversity. *Exceptional Children, 60*(6), 518-537.

Fullan, M., & Pomfret, A. (1977). Research on curriculum and instruction implementation. *Review of Educational Research, 47*(1), 335-397.

Heck, D., Chval, K., Weiss, I., & Ziebarth, S. W. (Eds.) (2012). *Approaches to studying the enacted mathematics curriculum.* Charlotte, NC: Information Age Publishing.

Hord, S. M., Stiegelbauer, S. M., Hall, G. E., & George, A. A. (2006). *Measuring implementation in schools: Innovation configurations.* Austin, TX: Southwest Educational Development Laboratory.

Huntley, M. A. (2009). Measuring curriculum implementation. *Journal for Research in Mathematics Education, 40*(4), 355-362.

Huntley, M. A., Rasmussen, C. L., Villarubi, R. S., Sangtong, J., & Fey, J. T. (2000). Effects of standards-based mathematics education: A study of the Core-Plus Mathematics Project algebra and functions strand. *Journal for Research in Mathematics Education, 31*(3), 328-361.

Jacobbe, T. (2009). The influence of standards-based curricula on questioning in the classroom. *NCSM Journal of Mathematics Education Leadership, 11*, 25-32.

Matsumura, L. C., Slater, S. C., Junker, B., Peterson, M., Boston, M., Steele, M., & Resnick, L. (2006). *Measuring reading comprehension and mathematics instruction in urban middle schools: A pilot study of the instructional quality assessment.* National Center for Research on Evaluation, Standards, and Student Testing (CRESST) Report #681.

Michlin, M., Seppanen, P., & Sheldon, T. (2001). *Charting a new course: A study of the adoption and implementation of standards-based mathematics curricula in eight Minnesota school districts.* Final report to scimath[MN], fall 2001.

Middleton, J. A. (1995). A study of intrinsic motivation in the mathematics classroom: A personal constructs approach. *Journal for Research in Mathematics Education, 26*(3), 254-279.

Moss, J., & Case, R. (1999). Developing children's understanding of the rational numbers: A new model and an experimental curriculum. *Journal for Research in Mathematics Education, 30*(2), 122-147.

National Council of Teachers of Mathematics. (1989). *Curriculum and evaluation standards for school mathematics.* Reston, VA: Author.

National Council of Teachers of Mathematics. (2000). *Principles and standards for school mathematics.* Reston, VA: Author.

National Research Council. (2004). *On evaluating curricular effectiveness: Judging the quality of K-12 mathematics evaluations*. Committee for a Review of the Evaluation Data on the Effectiveness of NSF-Supported and Commercially Generated Mathematics Curriculum Materials. J. Confrey & V. Stohl (Eds.). Mathematical Sciences Education Board, Center for Education, Division of Behavioral and Social Sciences and Education. Washington, DC: National Academies Press.

O'Donnell, C. L. (2008). Defining, conceptualizing, and measuring fidelity of implementation and its relationship to outcomes in K-12 curriculum intervention research. *Review of Educational Research, 78*(33), 33-84.

Papic, M. M., Mulligan, J. T., & Mitchelmore, M. C. (2011). Assessing the development of preschoolers' mathematical patterning. *Journal for Research in Mathematics Education, 42*(3), 237-268.

Porter, A., Schmidt, W. H., Floden, R. E., & Freeman, D. J. (1978). Practical significance in program evaluation. *American Educational Research Journal, 15*, 529-539.

Remillard, J. T. (2005). Examining key concepts in research on teachers' use of mathematics curricula. *Review of Educational Research, 75*(2), 211-246.

Remillard, J. T., & Heck, D. J. (2014). Conceptualizing the enacted curriculum in mathematics education. In D. R. Thompson & Z. Usiskin (Eds.), *Enacted mathematics curriculum: A conceptual framework and research needs* (pp. 121-148). Charlotte, NC: Information Age Publishing.

Reys, R. (2005). *Middle school mathematics study (MS)^2 assessing the impact of standards-based middle school mathematics curricula on teacher practices and student learning: Classroom observation protocol*. Columbia, MO: University of Missouri.

Reys, R. (2006). Assessing the impact of standards-based middle school mathematics curricula on student achievement and the classroom learning environment. Retrieved from http://mathcurriculumcenter.org/MS2_report.pdf

Schoen, H., Cebulla, K. J., Finn, K. F., & Fi, C. (2003). Teacher variables that relate to student achievement when using a standards-based curriculum. *Journal for Research in Mathematics Education, 34*(3), 228-259.

Seah, W. T., & Bishop, A. J. (2000). *Values in mathematics textbooks: A view through two Australasian regions*. Paper presented at the 81st Meeting of the American Educational Research Association, New Orleans, LA.

Shafer, M. C., Wagner, L. R., & Davis, J. (1997). *Teaching log*. (Mathematics in Context Longitudinal/Cross-Sectional Study Working Paper No. 5). Madison, WI: University of Wisconsin-Madison.

Silver, E. A. (1994). *Building capacity for mathematics instructional reform in urban middle schools: Contexts and challenges in the Quasar project*. Paper presented at the 1994 Annual meeting of the Educational Research Association, New Orleans, LA.

Smith, S. Z., & Smith, M. E. (2006). Assessing elementary understanding of multiplication concepts. *School Science and Mathematics, 106*(3), 140-149.

Stein, M. K., Remillard, J. T., & Smith, M. S. (2007). How curriculum influences student learning. In F. K. Lester (Ed.), *Second handbook of research on mathematics teaching and learning* (pp. 319-369). Greenwich, CT: Information Age Publishing.

Tarr, J. E., Reys, R. E., Reys, B. J., Chávez, Ó., Shih, J., & Osterlind, S. J. (2008). The impact of middle-grades mathematics curricula and the classroom learning environment on student achievement. *Journal for Research in Mathematics Education, 39*(3), 247-280.

Thompson, D. R., & Senk, S. L. (2001). The effects of curriculum on achievement in second-year algebra: The example of the University of Chicago School Mathematics Project. *Journal for Research in Mathematics Education, 32*(1), 58-84.

Weiss, I. R., Pasley, J. D., Smith, S., Banilower, E. R., & Heck, D. J. (2003). *Looking inside the classroom: A study of K-12 mathematics and science education in the United States.* Chapel Hill, NC: Horizon Research.

CHAPTER 6

CONCEPTUALIZING THE ENACTED CURRICULUM IN MATHEMATICS EDUCATION

Janine T. Remillard and Daniel J. Heck

Studying the enacted curriculum is facilitated when there exists a conceptual framework to provide a shared language for identifying and defining critical variables that make up and influence the enactment of mathematics curriculum in classrooms. This chapter proposes a conceptual framework initially developed at the Conference on Research on the Enacted Mathematics Curriculum. The framework is presented in two parts. The first part situates the enacted curriculum within a broad map of curriculum policy, design, and enactment, highlighting the variety of factors that influence the enacted curriculum. The second part delineates key components of the enacted mathematics curriculum. The chapter then illustrates how selected research studies map onto the two parts of the framework, highlighting the validity and viability of this conceptual framework for understanding current research and documenting where research is needed to help build knowledge within the field.

INTRODUCTION

A conceptual framework, in social science research, defines constructs under study and specifies their critical dimensions and the relationships among them that might be measured. The purpose of this chapter is to propose a conceptual framework that can guide future research on curriculum enactment and the enacted curriculum. The framework we present here identifies key features of the enacted curriculum situated within the broad system of curriculum policy, design, and enactment; it also identifies major contextual factors that influence various components of this system. Our goal in conceptualizing this complex set of relationships is to paint a clearer picture of the curriculum enactment process and identify important foci for research and policy analysis. In so doing, we hope to offer the field a shared language to guide research design and allow for comparison among, and combining of, studies to accumulate knowledge and generate new hypotheses.[1]

We begin by reviewing and comparing several conceptual frameworks used in large educational studies to ground our contribution in existing work and highlight places that our framework extends these perspectives. We then define "curriculum" as it is used in this chapter. We focus this definition on mathematics curriculum, acknowledging that the conceptualization of curriculum and the identification of relationships of interest have characteristics unique to mathematics education that affect how research is designed and conducted. Nevertheless, many of the relationships captured in the frameworks are drawn from research outside of mathematics.

The conceptual framework is then introduced in two parts. First, we situate the enacted curriculum within the broader system of curriculum policy, design, and enactment. Curriculum design, selection or designation, and enactment are all part of a complex process by which curricular intentions are conceptualized, adopted, substantiated, and enacted in the classroom (Ben-Peretz, 1990; Stein, Remillard, & Smith, 2007; Valverde, Bianchi, Wolfe, Schmidt, & Houang, 2002). We describe key components and relationships comprising this system that surround the enacted curriculum. Second, we elaborate elements that constitute the enacted mathematics curriculum. To illustrate analytic uses of the conceptual framework, we describe and situate a selected set of research studies that have examined various parts of the enacted mathematics curriculum over the past 20 years. This set of illustrative studies is not intended to be exhaustive. They were selected because they represent a range in scope and focus, grade range, and mathematical topics studied; they can be combined and compared as contributions to the body of knowledge charted in the conceptual framework. We conclude by proposing several ways that the conceptual framework can serve to further research on the enactment of mathematics curriculum.

BUILDING ON EXISTING CONCEPTUAL FRAMEWORKS

The framework presented here builds on and extends other approaches to conceptualizing and mapping the curriculum. Two of the most frequently cited examples of such frameworks are those used by Goodlad, Klein, and Tye (1979) in *The Study of Schooling* and by Schmidt et al. (1996) in TIMSS.[2] Both frameworks identify different levels of curriculum within an educational system. The *Study of Schooling* proposed five perspectives on curriculum: ideal, formal, instructional, operational, and experiential. The *ideal* reflects beliefs, opinions, and values of disciplinary and educational scholars, but most likely does not exist in reality. The *formal* curriculum refers to expectations for what should be taught and is represented in written statements, including state and district guidelines, department syllabi, commercially published materials, and local school documents. Goodlad et al.'s formal curriculum is akin to the *intended* curriculum defined by Schmidt et al., which includes intentions, aims, and goals located at the system level of schooling. Both of these frameworks distinguish this first perspective on curriculum from the *operational* (Goodlad et al., 1979) or *implemented* (Schmidt et al., 1996) curriculum, which is what actually happens in the classroom and exists within the domain of teachers' work. This level is also distinct from the curriculum *experienced* or *attained* by students. The framework employed by *The Study of Schooling* includes an additional curricular level that exists between the *formal* and *operational*: the *instructional* curriculum refers to teachers' intents and reflects their values and competencies.

Although these two frameworks use different terms, both seek to characterize a systemic perspective on what TIMSS defines as *educational opportunity* or "the configuration of social, political and pedagogical conditions to provide pupils chances to acquire knowledge, to develop skills and to form attitudes concerning school subjects" (Valverde et al., 2002, p. 6). As such, they conceptualize the relationship between curricular intents as they are specified in the policy realm and educational outcomes as measured by student experience and learning. Implicit in both frameworks is the assumption that curriculum exists in different forms on different planes. As it moves from the system to classroom to student level, curriculum not only becomes more specified, but is also reformulated in the hands of different actors. In developing the framework presented here, we have taken these assumptions as a starting place.

A central feature of interest in the framework we present is the place of instructional materials for the mathematics classroom. The two frameworks on which we are building differ in how they position instructional materials. Goodlad and colleagues (1979) identify instructional materials as part of the formal curriculum because they are written statements of intent. Schmidt and colleagues (1996) frame the intended curriculum more narrowly, as aims and goals, and view instructional materials or textbooks as mediators between the intended and implemented. Because the intended curriculum rarely specifies what must take place to enact instructional activities, "textbooks are written to serve teachers and students in this way—to work on their behalf as the links between the ideas presented in the intended curriculum and the very different world of the classroom" (Valverde et al., 2002, p. 55).

Two additional frameworks that take written instructional materials as a starting place for analysis have also informed our work. The Math Task Framework used in the QUASAR study (cf. Stein, Grover, & Henningsen, 1976) highlights three levels of mathematical tasks that must be considered in order to understand the influences that curriculum, teachers, and the social dynamics of the classroom have on student learning. Stein et al. differentiate between mathematical tasks "as represented in curricular/instructional materials," tasks "as set up by the teacher in the classroom," and tasks "as implemented by students in the classroom." Like the two previous frameworks, the Math Task Framework acknowledges that tasks change in their features and, frequently, in their level of cognitive demand as they move through this process. Although the Math Task framework does not address system-level influences on the curriculum, it captures in greater detail than the Goodlad or Schmidt frameworks the roles that classroom norms and interactions, and students' habits and dispositions, play in shaping the enacted curriculum. Research findings from the QUASAR study that employed this framework found that shifts in tasks that reduced their cognitive demand resulted in lower levels of student learning. Stein, Remillard, and Smith (2007) adapted the Math Task Framework to present findings from a review of research on the influence of curriculum on student learning. Their framework incorporates three temporal phases of curriculum use: the written curriculum (referring to what appears on the printed page in instructional materials), the intended curriculum (referring to a teacher's plans for instruction), and the enacted curriculum (referring to the implementation of curricular-based tasks in the classroom). A primary focus and contribution of the original and adapted Math Task Frameworks is to offer explanations for how these shifts take place; in this spirit, both frameworks include factors that influence these "transformations," as they are referred to by Stein et al. (2007).

The framework we present here adopts many of the points of convergence of these previously developed frameworks in seeking to capture the different levels, perspectives, or instantiations of curriculum across domains of the educational system. In addition, we seek to place a magnifying glass over two components of the system, the formal or official curriculum and the enacted curriculum, in order to reveal their interworkings. Our analysis of contemporary policy, particularly in the United States, suggests that the educational system has multiple levels, and that states, districts, and schools play varied roles in establishing the intended curriculum. We seek to articulate and explain these details. Looking across studies of mathematics classroom practice, we seek to elaborate dimensions that comprise the enacted curriculum. Given the differing ways instructional materials are conceptualized in existing frameworks, we seek to articulate the roles these resources play in both components.

Definitions

We define *mathematics curriculum* as a plan for the experiences that learners encounter and the actual experiences that are designed to help them reach specified learning goals for mathematics. We use the term *plan for the experiences* to signal that curriculum is more than the specification of learning goals or topics; it includes a plan for the learning experience students will encounter. It also includes, in the case of enacted curriculum, the *actual experiences* students encounter. This particular definition of curriculum is informed by the original Latin meaning of curriculum—a course, specifically the course or track of a chariot race. In this sense, it is more than the end point or goal, and includes the path traveled. In the definition for mathematics curriculum, we have specified *for mathematics* for several reasons. First, as mathematics educators, we acknowledge that our focus is limited to the mathematics curriculum, although our framework has been influenced by research outside of mathematics. That said, we believe this framework would benefit from scrutiny and feedback from scholars in other fields included in the school curriculum. Second, the content of mathematics has its own conceptual and scientific research base; this base has its own structure and customs that influence conceptualizations in the field. Third, we recognize that the school curriculum, even when focusing on mathematics, is not and should not be attentive to mathematics learning goals alone. As social institutions, schools have multiple purposes and responsibilities.

SITUATING THE ENACTED CURRICULUM IN A POLICY, DESIGN, AND ENACTMENT SYSTEM

In this section, we situate the enacted curriculum within a broad system of curriculum policy, design, and enactment that captures the relationships among the different instantiations of curriculum. This system is represented by the framework shown in Figure 6.1. The rectangles in the framework refer to curriculum variables, or curriculum in its many forms. These variables include curriculum plans (i.e., the designated curriculum, the curriculum outlined in instructional materials, and the teacher intended curriculum), curricular goals, and the enacted curriculum. We use an oval to show student learning because it is not a curriculum variable, but a curriculum outcome. The variables are given different shades to distinguish their role in the system. The absence of shading indicates curricular goals. Light and dark shading indicate different types of planned or experienced curriculum, with light shading employed to designate written curriculum, and dark shading used to specify classroom-level and teacher-level enactments—the curriculum intended or planned by the teacher and the curriculum actually enacted in the classroom. The framework also identifies a number of mediating factors that influence different parts of the system. These factors are identified in rounded rectangles around the perimeter of the framework.

In the framework, we distinguish between curricular elements that are *officially* sanctioned and those not sanctioned but operationalized through practice. The elements of the curriculum are specified or adopted by authorizing entities for a particular system or school and include curricular goals, content of assessments, and the designated curriculum. The *operational* elements are those that actually occur in practice. The arrows in the diagram signify paths of likely influence, suggesting questions or hypotheses that could be the focus of research. Some of these paths have been empirically explored; others are speculative. It is worth noting that these paths of influence represent points of connection across different levels of, and players in, the system. In some cases, they highlight places where a hand-off of sorts occurs; it is at these points that interpretations and translations necessarily occur. These pathways, their strength, and their character are all potentially important objects of study to further our understanding of the enacted mathematics curriculum.

The Official Curriculum

The components of the official curriculum are specified and authorized within a school or school system. The key components of the official curriculum are (a) curricular goals, (b) content of consequential assess-

Conceptualizing the Enacted Curriculum in Mathematics Education

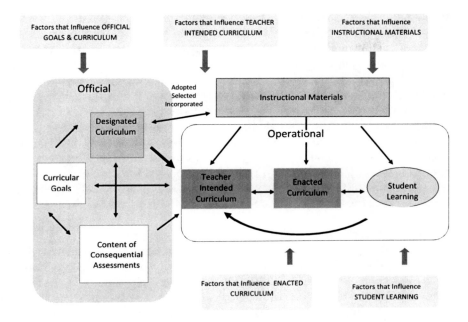

Figure 6.1. Framework of the curriculum policy, design, and enactment system.

ments, and (c) the designated curriculum. These components refer to what McLaughlin (1990) might call the policy structure within a system and take into account national and state, as well as local, policy.

Curricular Goals

Curricular goals are the specified learning outcomes often set or adopted by a national body, state or province, school system, or school. These are statements of what students should ideally learn as a result of instruction. Researchers have called the set of such specifications the formal or intended curriculum. In the United States, the contemporary term for such goals is *standards* at the state or national level, although the terms *learning outcomes* and *objectives* are often used in the United States; *curriculum framework* is a term frequently used in a number of countries to refer to specified learning goals. We call them *curricular goals* because they are aimed at shaping the curriculum, but they are not the curriculum because they do not specify plans for enactment.

Content of Assessment

Curriculum is also shaped by the content of official assessments that are consequential for students or the school, which some researchers have termed the *tested curriculum*. These assessments are often used to track progress or measure the instructional quality of a school or individual students' achievement. Some of these assessments reflect the stated curricular goals and are designed to assess students' attainment of these goals. It is important to note, however, that any curriculum-referenced assessment represents one particular instantiation of these goals, reflecting the interpretation embraced by those constructing the test as well as technical and logistical constraints of the particular test. Moreover, the content of consequential assessments can influence official curricular goals, so a bidirectional arrow between assessments and goals is shown in the framework. Knowledge and skills that can be readily measured using consequential tests are likely to be incorporated into official curricular goals, both as they are initially stated and as they are clarified over time.

The Designated Curriculum

We use the term *designated curriculum* to refer to the set of instructional plans specified by an authorized, governing entity, be it a district, school, or consortium of schools. The designated curriculum is generally informed by the official curricular goals and is intended to offer guidance toward addressing these goals. However, during periods of high accountability, the designated curriculum is even more likely to be influenced by the content of assessments that students take and for which schools and districts are held accountable. The designated curriculum can vary in its form and degree of specificity. In some settings, the designated curriculum is simply an adopted mathematics textbook, in which case it is identical to what many call the written curriculum; in others, the designated curriculum is comprised of a host of assembled packages of materials, instructional resources, and structuring guidelines designed to shape the content, pacing, and often the processes and tools of mathematics instruction. Instructional materials, shown outside the official domain in the framework, become part of the official curriculum when they are designated as an adopted or required resource.

The designated curriculum is the element of the official curriculum intended to have the most direct influence on the enacted curriculum. As previously mentioned, it generally offers instructional specificity that curricular goals and assessment items do not. In many cases, the designated curriculum includes adopted instructional materials. In some cases, it comprises only a selected set of materials.

The inclusion of the designated curriculum and the placement of instructional materials outside of the official curriculum are two important elements of this framework that contribute to understanding of curriculum design and enactment. Most curriculum frameworks specify written curriculum goals or objectives as the official or intended curriculum and curriculum materials as the written curriculum (see Cal & Thompson, this volume for details). Neither of these representations alone captures the complexity or the specificity provided by what we refer to as the designated curriculum. The designated curriculum is an aggregate of curricular goals, instructional resources, and implementation guidance tailored to the local context. Even though instructional materials offer written curricular guidance, they do not become part of the officially sanctioned curriculum unless they are incorporated into the designated curriculum. In our view, the designated curriculum takes the place of the textbook in the framework used in TIMSS as the mediator between intention and implementation (Valverde et al., 2002). That said, the designated curriculum typically includes instructional materials.

The Operational Curriculum

When put in the hands of teachers, the official curriculum, as conveyed through the designated curriculum, begins to change form, moving from descriptions of instructional activities toward actual classroom interaction. We refer to these components of the curriculum as *operational* because they include the transformations that occur through the enactment process. Our understanding of the enactment process is informed by decades of study of the on-the-ground implementation of instructional innovations, including curriculum. The *Rand Change Agent Study* (summarized and discussed by McLaughlin, 1990), in confirming that implementation involves a process in which both the innovation and the local context are mutually adapted, found that local factors, especially on-the-ground actors, tend to dominate outcomes. McLaughlin (1990) reminds us that, although the policy structure within a system is most visible to outsiders, to local actors, especially teachers, these formal structures are simply part of a broader set of contextual pressures and influences. Other researchers have shed light on the interpretive and meaning-making processes involved in bringing new ideas and directives into practice (e.g., Coburn, 2006; Cohen & Hill, 2001; McLaughlin, 1990). In the proposed framework, the operational curriculum includes (a) the teacher intended curriculum, (b) the enacted curriculum, and (c) student learning (sometimes called the *learned curriculum*).

Teacher Intended Curriculum

As teachers draw on the designated curriculum along with other resources to design instruction, they create what we call the *teacher intended curriculum;* it includes the interpretations and decisions made by the teacher in order to envision and plan instruction. This form of curriculum, like those previously discussed, remains composed of projected possibilities for classroom instruction. At the same time, it has more texture and detail than the previous forms; it is designed for specific students at a particular moment in time and is ready to come alive in the classroom. This form of curriculum is also the most difficult to access to study because it exists in its most detailed state in the teacher's mind. Lesson plans can be used as artifacts of a teacher's plans, but they are unlikely to capture much of the multi-layered nature of a planned lesson as fully imagined by the teacher. The difference between the designated and teacher-intended forms of curriculum is akin to the difference between a script of a play and each scene as conceived by the director.

Teachers are the primary intended audience of the designated curriculum. There is considerable evidence that the designated curriculum—including district-level specifications of learning expectations, adopted instructional materials, and pacing guides—serves as mathematics teachers' primary instructional resource and guide for what mathematics content to address and, to a lesser degree, how to address it (Sosniak & Stodolsky, 1993; Weiss, Pasley, Smith, Banilower, & Heck, 2003). Nevertheless, the two other components of the official domain—adopted curricular goals and the content of consequential assessments—are likely to influence teachers' curricular decisions directly (as well as indirectly through the designated curriculum). Differences in the likely intensity of influence are represented by the thickness of arrows in the framework in Figure 6.1 but need to be developed empirically for each setting or circumstance.

The importance of the conceptual distinction between the teacher intended or planned curriculum, the designated curriculum, and the enacted curriculum is illustrated by research that highlights teachers' interpretive activities when using curriculum resources to plan instruction (Lloyd, 1999; Remillard, 1999) and the unfolding nature of the enacted curriculum (Stein, Grover, & Henningsen, 1996). Many models of the curriculum implementation process go straight from the written (or designated curriculum) to the enacted curriculum. By including the teacher intended curriculum in this framework, we are returning to Goodlad and colleagues' (1979) inclusion of the instructional level in shaping the curriculum and acknowledging that the operational curriculum has at least two phases: the first occurs when the teacher transforms the designated curriculum into plans for instruction; the second occurs when those plans are enacted in the classroom.

Enacted Curriculum

The *enacted curriculum*, the interactions between teachers and students around the tasks of each lesson and accumulated lessons in a unit of instruction, is analogous to the performance of a play, complete with the idiosyncrasies and unpredictable elements of live performance. The enacted curriculum cannot be scripted because the enactment itself requires teachers to respond in the moment to the events in the lesson. That said, the enacted curriculum is directly influenced by the teacher intended curriculum, the instructional resources being employed, the students, a host of contextual factors, and the teacher's ongoing responses to these variables. The double arrow between the teacher intended and the enacted curriculum signals that the teacher intended curriculum can be dynamic; teachers are likely to make revisions in their plans as the curriculum is being enacted.

The notion that the enacted curriculum (also referred to as the implemented curriculum or classroom instruction) is distinct from the written or planned curriculum has been embraced by many curriculum scholars. In a review of research on curriculum implementation, Snyder, Bolin, and Zumwalt (1992) assert that "a major modification to the curriculum implementation literature was made by researchers who discovered that, in reality, curriculum was never really implemented as planned, but rather adapted by local users" (p. 428). From one perspective, these adaptations are viewed as departures from the established curriculum. From another perspective (an enactment perspective according to Snyder et al., 1992), the enacted curriculum is, by definition, an emergent, jointly constructed entity.

The enacted curriculum is the element of the entire curriculum framework that has the greatest impact on student learning. Its proximity to classroom learning makes understanding it and developing ways to reliably measure its dimensions critically important (Raudenbush, 2008). We discuss how the enacted curriculum is conceptualized in terms of its primary components in a later section.

Student Learning

The variable often of most interest in studies of the curriculum implementation process is student learning—the residue (to use a term introduced by Hiebert et al., 1997) that remains with students as a result of participation in the enacted curriculum. We choose to define learning broadly to acknowledge the multiple ways that learning is conceptualized and to emphasize that the outcomes of mathematics learning are greater than the content knowledge and skills typically assessed on tests. Knowledge, skills, and practices are critical parts of student learning noted in the framework and are assessed as products of the enacted curriculum.

Some products of the enacted curriculum are not explicitly or intentionally taught. We take learning to include the influences of instructional experiences on attitudes, interests, perceptions, and identities that have substantial consequences for students' future trajectories. Scholars use the term "hidden curriculum" to refer to messages communicated to students through instruction and the context of instruction (Dreeben, 1968; Jackson, 1968; Snyder, 1971). These messages usually convey perceptions about the subject matter, what it means to know it, and about the learners themselves, including how students ought to behave in math class. A number of studies illustrate what students learn about mathematics beyond the explicit curriculum. Verschaffel, De Corte, and Lasure's (1994) study of fifth grade students illustrates what years of mathematics instruction can teach children about what answers should look like in a math classroom. When given tasks that were situated in real-world contexts, students tended to ignore the contexts and offered answers that were mathematically correct, but made no sense in the given situation.

We see student learning as being primarily shaped by the enacted curriculum and the instructional materials. The relationship between student learning and the enacted curriculum is depicted as bidirectional because as students interact in the classroom with the tasks, the materials, each other, and the teacher, they co-create the enacted curriculum.

Contextual and Influencing Factors

Drawing on a large body of existing research, we understand the elements of both the official curriculum and the operational curriculum to be influenced by a number of mediating factors. These factors may be social, political, structural, or cognitive and they may exercise their influence through formal and informal policy, collectively shared opinions or perspectives, institutional constraints or supports, and individual human capacity. These factors are situated around the perimeter of the framework and can easily be interpreted as external forces pressing in on the curriculum system. We view these factors as natural and integral to the system of curriculum policy, design, and enactment, rather than separate from it. Further, the way these factors exercise influence within the system varies and, in some cases, needs further exploration and elaboration. Thus, we conceptualize them as mediating variables and potential objects of study.

Our framework identifies a set of possible influencing factors according to the objects of influence, that is, factors that influence official goals and curriculum, instructional materials, teacher intended curriculum, enacted curriculum, and student learning. The following factors are those most commonly cited in the literature, but are not intended to represent an exhaustive list. Further, their mode of influence can be explicit and direct, or intricate and subtle. Influencing factors may reinforce or conflict with one another and their resulting impact on the curriculum is not straightforward. It is also important to note that, as the pathways in the framework suggest, many key variables are expected to have a role in mediating other variables.

Some of the factors that influence official goals and curriculum include: perceived and expressed needs of society; advancements in the fields of mathematics, learning, educational practice, assessment, and technology; policies, accepted assessment practices and constraints; values and beliefs about mathematics and the goals of education as held publicly and by individuals and groups yielding power. Factors that are likely to influence the content and design of instructional materials include: market forces; the views of professional societies, research on learning, official curricular goals and the factors that influence them; and curriculum developers' visions. Factors that tend to influence the teacher intended curriculum include: teacher knowledge, beliefs, and practices about mathematics, pedagogy, learning, and curriculum resources; access to resources and support; understanding of particular students' needs; and the local context. Factors that influence the enacted curriculum include: teacher and student knowledge, beliefs, and practices; access to resources, such as instructional technologies; and contextual opportunities and constraints. Factors that may influence student learning include: students' prior knowledge; identity, attitudes, motivation; home support and beliefs of others in the home; peers; and comfort in the classroom and with the content.

RESEARCH THAT EXAMINES THE CURRICULUM POLICY, DESIGN, AND ENACTMENT SYSTEM

The conceptual framework proposed here to characterize the curriculum policy, design, and enactment system provides a heuristic for designing and situating studies that examine relationships among elements of curriculum and influences on them. In addition to mapping the terrain and the relationships within it, the framework can assist researchers in taking stock of the knowledge base on the enacted mathematics curriculum. In the discussion that follows, we use several historically prominent as well as some recently published studies to illustrate the use of the framework as a conceptual map and begin to summarize existing knowledge in the field.

Studies of the enacted curriculum within the curriculum policy, design, and enactment system have focused on different kinds of questions and on different components of the system. Some studies seek to shed light on the system as a whole, identifying key variables and illustrating how they interact. These studies address questions such as:

- How do curriculum-related policies and decisions at various levels of the system interact with one another?
- What types of curriculum-related policies are predictive of the mathematics curriculum that is enacted in classrooms?
- What variables influence the relationship between the content assessed and the content taught in the classroom? And, ultimately, what factors most influence student outcomes?

Within these studies, some connect the enacted curriculum to other forms of curriculum that are "upstream" in the system and serve as contributing factors. For example, in a study of elementary and middle grades classrooms across eleven states in the United States, the Council of Chief State School Officers (CCSSO), Wisconsin Center for Education Research, and Eleven State Collaborative (2000) investigated the alignment between the enacted curriculum as reported by teachers and key elements of the official curriculum, including state standards, the content emphasis and student learning expectations communicated through state-level curriculum goals, and the content of high-stakes assessments. The enacted curriculum was measured using a survey of enacted curriculum, on which teachers reported content covered and their student learning expectations for specific mathematical topics.

Findings from these teacher self-report measures suggested that state and national standards had a strong influence on the enacted curriculum as reported by teachers. Teachers generally indicated that they covered the content specified in recommended standards, but reported emphasis on student learning expectations with much greater variability. Teachers' self-reports of the enacted curriculum in their classrooms more closely reflected the content emphasis seen in standards on high learning expectations, such as problem solving, conceptual development, and applications, compared to the content emphasis of high-stakes tests, which tended to emphasize lower learning expectations such as memorization and procedural skill.

A study by Tarr and colleagues (2008) provides another example of an effort to examine the relationships among different pieces of the curriculum policy, design, and enactment system. The study compared mathematics instruction and student learning in 10 middle schools in the U.S., focusing the analysis on relationships among instructional materials and their use, the enacted curriculum, and student learning outcomes. Half of the teachers in the study were using one of three mathematics programs designed to follow the recommendations in the *Curriculum and Evaluation Standards for School Mathematics* (National Council of Teachers of Mathematics, 1989). The other half were using one of five different conventional programs. The researchers found significant variation in the enacted curriculum, regardless of the curriculum program being used. Not all teachers using *Standards*-based materials enacted *Standards*-based instruction; however, none of the teachers using conventional materials enacted instruction that could be considered *Standards*-based. In other words, the instructional materials had some influence on the enacted curriculum, but they were certainly not the only factor.

The element of the curriculum policy, design, and enactment system that is most frequently studied is student outcomes. Consistent with characterizations of student learning as the achieved curriculum or the learned curriculum, the proposed conceptual framework identifies student learning as a form of curriculum that is "downstream" from the enacted curriculum; however, student outcomes are defined and measured differently by different researchers. The Tarr et al. (2008) study used two different instruments for measuring student achievement: the *Terra Nova Survey* (published by CTB-McGraw Hill), which measures knowledge and application of procedures and concepts; and the *Balanced Assessment in Mathematics* (BAM) (also published at the time of the study by CTB-McGraw Hill), which measures problem solving and conceptual understanding. Students in the study showed no significant difference in scores on the *Terra Nova*, regardless of other "up-stream" factors they experienced. Yet, the students who experienced *Standards*-based instruction for at least two years demonstrated higher levels of achievement on the BAM than those who did not.

Tarr et al's. (2008) finding that the quality of the enacted curriculum seems to have the greatest influence on student learning comports with other research findings. The QUASAR study, for example, assessed student performance using project-designed measures of reasoning and problem solving. Students in classrooms where teachers not only posed high cognitive demand tasks but also maintained the cognitive demand throughout the lesson performed better on the assessments (Stein & Lane, 1996).

Other variables used to assess the quality of student outcomes from the enacted curriculum include students' views of math (Boaler & Staples, 2008), their sense of themselves as mathematical (Boaler & Greeno, 2000), and their ability to contribute to classroom discourse (Herbel-Eisenmann & Otten, 2011). These and other studies are predicated on the perspective shared by many scholars that developing a positive mathematical disposition and the ability to engage in mathematical practice are valuable mathematical goals for students, even though they are not measured by achievement tests (National Research Council, 2001).

In addition to research that situates the enacted curriculum within the curriculum policy, design, and enactment system, research on mathematics curriculum may attend specifically to the nature and quality of the enacted curriculum and the factors that influence it. The following section describes an expanded view of this central component of the conceptual framework, identifying key constructs that make up the enacted curriculum.

CONCEPTUALIZING THE ENACTED CURRICULUM

We define the *enacted curriculum* as the interactions between teachers and students around the mathematical tasks of each lesson and accumulated lessons in a unit of instruction. As described previously, the enacted curriculum is the aspect of the curriculum closest in proximity, and we argue in influence, to student learning. We do not suggest that instructional resources, the designated curriculum, or curricular goals do not influence student learning; their influence, however, is dependent on how they figure into the enacted curriculum. For this reason, understanding the enacted curriculum and developing ways to study it are critically important.

Despite its importance, the enacted curriculum has been difficult to measure and study. One challenge in measuring the components of the enacted curriculum is the complexity of classroom activity. The number of potential features of interest is great and many features are difficult to define and measure. Indeed, scholars studying the enacted mathematics curriculum have identified a wide range of components of interest to research.

Our conceptualization of the enacted mathematics curriculum is informed by our reading of theoretical and empirical literature and our own research. We identify four key dimensions of the enacted curriculum: (a) the mathematics; (b) instructional interactions and the norms that govern them; (c) the teacher's pedagogical moves; and d) the use of resources and tools. We think of these dimensions as a set of broad categories that interact with one another to form the primary components of the enacted curriculum and that merit further study. In Figure 6.2, we propose a model for understanding the relationships among these dimensions, but acknowledge that continued research is needed to refine it.

The Mathematics

The *mathematics* refers to the content and nature of the mathematics that is emphasized and valued in the class. It includes the particular content ideas that are covered and how they are represented and engaged. Examinations of the mathematics might span the range of topics and

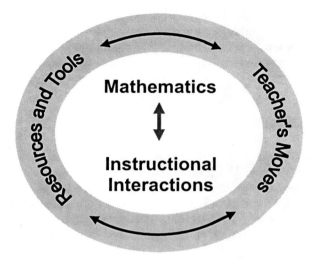

Figure 6.2. Model of the dimensions of the enacted curriculum.

concepts covered, or pinpoint specific topics and concepts. Moreover, research might investigate what it means to know and do mathematics in the classroom(s) being studied, attending to the mathematical practices by which students encounter and make meaning of ideas and skills, and the expectations for students to engage in those practices.

Researchers studying the mathematics of the enacted curriculum focus on different aspects of this dimension. Some focus on the content of mathematics covered as well as its nature and quality. Tarr and colleagues (2008) as well as CCSSO et al. (2000) examined the mathematics content of the enacted curriculum, in part by the level of emphasis on various mathematics topics across the course of a school year. Another way to investigate mathematics content as a part of the enacted curriculum is evident in the TIMSS videotape studies, which compared selected eighth-grade lessons in Japan, Germany, and the United States (Stigler, Gonzales, Kawanaka, Knoll, & Serrano, 1999), and later investigated nationally representative samples of eighth-grade lessons in the United States and six higher-achieving nations (Australia, Czech Republic, Hong Kong, Japan, Netherlands, Switzerland) (Hiebert et al., 2003). These studies examined and compared the means by which mathematics content was presented to students, considering features like demonstration, practice, recall of concepts, conceptual connections, and proof. Studies by Boaler and Staples (2008), Eisenmann and Even (2009), and Stein, Grover, and Henningsen (1996), as well as the CCSSO-led study (2000), included another facet of consideration with respect to mathematics content, focusing on the level of cognitive expectations for student learning of mathematics ideas. Stein et al., for example, analyzed the nature and quality of the mathematical tasks, focusing, in particular, on their cognitive demand.

Instructional Interactions

Instructional interactions in Figure 6.2 refer to the interactions that take place among students, the teacher, the mathematical tasks, and the available tools together with the norms that govern them. These interactions are depicted by Cohen and Ball (1999) in a triangle that positions interactions among teachers and students around educational material at the heart of instruction. Other researchers have focused specifically on the nature of the discourse in the classroom as a critical component of the enacted mathematics curriculum (Herbel-Eisenmann & Otten, 2011; Wood, Nelson, & Warfield, 2001). These interactions reflect and shape the way mathematics is positioned and represented in the classroom and influence it, in turn influencing the nature of the mathematics dimension of the enacted curriculum.

Given the central role that instructional interactions play in the enacted curriculum, a number of researchers have attended to norms that govern these interactions. Researchers, such as Hiebert et al. (1997) and Yackel and Cobb (1996), argue that norms established over time in a classroom, or any other instructional setting, shape the expectations for interactions among the group of students and foster the development of particular routines of practice. These norms shape what counts as valued mathematical work in the classroom (Franke, Kazemi, & Battey, 2007).

A number of different approaches to measuring or describing instructional interactions and supporting norms, either between teacher and students or among students, have been employed in studies of the enacted curriculum. For instance, Goos (2004) investigated grade 11 and 12 high school classrooms in an independent school in Australia. Over two years, observations and video- and audio-recordings, along with stimulated-recall interviews of the teacher and students, were used to characterize the nature, purpose, and interpretations of classroom interactions. The study illustrated specific ways the teacher's shaping of classroom interactions resulted in the development of a community of mathematical inquiry and sense-making in the classroom.

Yackel and Cobb (1996) argued for a position on classroom norms that places particular emphasis on how the teacher and students engage in *mathematical* practice. They called these norms of engagement *sociomathe-*

matical norms. Yackel and Cobb claimed that sociomathematical norms are based on the mathematical integrity of social practice in a particular context and include "what counts as mathematically different, mathematically sophisticated, mathematically efficient, and mathematically elegant" (p. 461). For example, agreement on what counts as an acceptable mathematical explanation is a sociomathematical norm, which may differ widely from one classroom to another, or even from one topic to another.

Each of the previous examples of research analyzing instructional norms and interactions in the classroom highlights the role that interactions around the mathematics tasks play in shaping how mathematics is represented, what it means to be mathematical, and what type of mathematical knowledge is valued. The term *sociomathematical norms* captures how closely intertwined these dimensions are. For this reason, mathematics and instructional interactions are at the center of the model and sit in relation to one another. Close examinations of classroom interactions also reveal the critical role teachers play in shaping these interactions and establishing normative practices. Other research suggests that the tools and resources available and taken up are also integral to the curriculum as enacted. Because of their shaping and constraining roles, these two dimensions, elaborated in the following sections, are placed around the outer ring of the model shown in Figure 6.2.

Teacher Pedagogical Moves

In Figure 6.2, *teacher pedagogical moves* refer to a teacher's actions, both intentional and unintentional, that shape what mathematics is covered and how it is represented and investigated. Teacher moves also influence how the classroom interactions are structured, the kinds of interactions that are valued, and which tools and resources are used during instruction. The primary focus of research on teacher's pedagogical moves is the collection of actions teachers take to structure and manage students' engagement with mathematical tasks in the classroom. Researchers focus on how teachers shape the mathematical discourse in order to extend students' mathematical thinking through facilitating talk, the use of gestures and other actions, and the strategic use of mathematical models and representations. In a review of research on the teacher's role in mathematics classroom discourse, Walshaw and Anthony (2008) identify three distinct roles that teachers enact to shape classroom discourse: (a) identifying and drawing out specific mathematical ideas, (b) fine-tuning the mathematical language and conventions used, and (c) shaping mathematical argumentation as it develops. We offer a few illustrative examples of such studies.

Herbel-Eisenmann and Otten (2011) conducted an intensive study of classroom discourse using thematic analysis from systemic functional linguistics. Their investigation of language use in two middle-grades classrooms in the U.S. revealed notable contrasts in how teachers' pedagogical moves differentially encouraged students to share their mathematical thinking or to attend to other students' thinking. Shein (2012) similarly attended to teacher pedagogical moves in a study of a grade 5 teacher's classroom in the United States, investigating the teacher's use of mathematically relevant gestures in her instruction with students who were not native speakers of English, the primary language used in the classroom. Shein reported that the teacher's use of gestures bolstered these students' participation in the classroom to examine and repair errors in problem solving. Attention to how teacher actions highlight key mathematical ideas, including ideas embedded in student thinking, was also evident in the study of two grade 7 classrooms in Israel by Eisenmann and Even (2009), who focused on the teacher's task and question selections.

Tools and Resources

Tools and resources in Figure 6.2 refer to physical, technological, linguistic, and cognitive tools that might be employed during instruction by both teacher and students. Tools include instructional resources, like texts, as well as concrete resources, like calculators, computers, and manipulatives, in addition to mathematical representations, algorithms, and mnemonic strategies. Tools are often introduced into the classroom through teachers' moves and influence how the mathematics is represented and engaged as well as the nature of the classroom interactions. At the same time, classroom interactions and norms also shape the way tools are employed.

Hollebrands's (2007) study of the use of dynamic geometry software in a secondary classroom in the U.S. provides one illustration of how tools can shape instructional interactions. This technological tool was incorporated regularly into instructional interactions so that students used the software frequently as they worked on mathematics tasks in the classroom. Hollebrands found that students' engagement with the software during classroom instructional interactions aligned with the nature of their problem solving on independent assessment tasks completed apart from regular instruction. Students' interpretations of the mathematical meaning and implications of their own work during these assessment tasks corresponded closely to the ways that they used the software to complete tasks.

Instructional materials provide another example of a resource found to shape the nature of classroom interactions. Many researchers have examined how instructional materials are used during curriculum enactment. There is general agreement that these resources have an influence on both mathematical content and pedagogical emphasis of the enacted curriculum, but this influence is moderated by other critical variables, including the content of the official curriculum, the design of the materials, teacher and student characteristics, and features in the context (see, e.g., Gueudet, Pepin, & Trouche, 2011; Gueudet & Trouche, 2009; Remillard & Bryans, 2004; Stein & Kaufman, 2010; Stein, Remillard, & Smith, 2007).

RESEARCH ON FACTORS THAT INFLUENCE THE ENACTED CURRICULUM

Thus far, our efforts to map existing research onto the curriculum policy, design, and enactment system have focused, first, on studies that delineate the elements of the system and the relationships among them and, second, on studies that illuminate the critical components of the enacted curriculum. There is also considerable interest in determining factors that shape the enacted curriculum because of its value in predicting student outcomes. We now describe several such studies.

Teachers and students surface in the research as having substantial influence on the enacted curriculum. Research demonstrates that teachers using the same instructional materials, even in the same school context, use the materials in different ways and enact markedly dissimilar instruction (Stein, Remillard, & Smith, 2007). Some teacher factors identified in the literature include mathematical knowledge (e.g., Hill & Charalambos, 2012), beliefs about teaching and learning mathematics (e.g., Lloyd, 1999; Sleep & Eskelson, 2012), and their comfort with the curriculum (e.g., Remillard & Bryans, 2004).

Students have been found to influence the enacted curriculum in two important ways. First, they influence teachers' perceptions of what is appropriate and necessary when designing instruction; second, they participate with teachers in shaping the enacted curriculum. Eisenmann and Even (2009) studied a single teacher in Israel teaching the same unit from the same instructional program to two seventh-grade classes in different schools. The study identified school-level expectations for teachers and students, discipline, and student engagement in the classrooms as key influences on choices the teacher made about what specific tasks to use with students and how to engage students with the mathematics content of those tasks. Stein, Grover, and Henningsen's (1996) examination of how cognitive demand tended to decrease during instruction, often when students had difficulty or resisted challenging tasks, also reveals roles that students play in shaping the enacted curriculum.

Available resources influence the enacted curriculum through planning and during the enactment process. Hollebrands's (2007) study illustrates the ways that consistent use of a software program can lead to kinds of problem solving and mathematical thinking unlikely without these tools. Although teachers' perceptions moderate their use of instructional materials, the content and quality of these materials certainly matter. Stein and Kaufman's (2010) study examined instructional practices of 48 teachers, spread across the elementary grades, in two school districts in the United States. The study investigated the relationship between the particular aspects of the enacted mathematics curriculum and teachers' mathematical knowledge, features of the district contexts, and their use of the guidance and resources provided in the instructional materials. One key finding of this study was that when teachers' lesson preparation included consideration of key mathematics ideas, their enactment of lessons was more likely to maintain high cognitive demand. Moreover, a comparative analysis of the two curriculum programs (Stein & Kim, 2009) indicated that the program more closely associated with high cognitive demand during instruction also placed greater emphasis on providing teachers with more extensive supports.

A final factor moderating the enacted curriculum discussed in the literature is the local, school, or regional context. Notable contextual factors include local norms, politics, and accepted practices, as well as structures in the environment and regional or national policy. The *Inside the Classroom* study by Weiss et al. (2003) in the U.S. provides insight into the role that policy plays in shaping the enacted curriculum. Researchers observed a nationally representative sample of 186 classrooms spanning grades K-12. Along with attention to the nature and quality of the enacted curriculum, the study investigated influences on teachers' intentions related to the content and pedagogy of mathematics lessons, and found that state or district standards, curriculum materials, and assessments used in accountability systems were the most commonly identified influences on lesson content. By contrast, the most frequently identified influences on lesson pedagogy were teachers' knowledge, beliefs, and experiences, followed by curriculum materials and characteristics of students.

The role that local norms and structures play in shaping the enacted curriculum is illustrated by a study of an eighth-grade algebra teacher by Herbel-Eisenmann, Lubienski, and Id-Deen (2006). The particular teacher taught two distinct versions of the course, using contrasting textbooks—one followed the goals of the NCTM *Standards*; the other followed a conventional approach. Students self-selected into their algebra course. The researchers were surprised to find that the same teacher used very different instructional strategies and established contrasting learning environments in the two sections of the course. These differences in enacted curriculum were attributed only partly to the different instructional programs being used. Interviews of the teacher and student surveys indicated that the teacher's instructional decisions were also influenced by parents' and students' expectations and the local politics within the district around educational reform.

CONSIDERING TIME AND ORGANIZATIONAL LEVELS

Despite our efforts to situate the enacted curriculum within a complex context of policy and social factors that influence it, there are also ways that the framework represented in Figure 6.1 is oversimplified in its representation of the mathematics curriculum. We raise two critical considerations briefly here that need to be considered in research design: time spans and organizational levels. Both require the researcher to consider the way the variables in the framework would be best understood as longitudinal, and nested or layered phenomena.

Considerations of time span emerge when we recognize that curriculum can be examined by attending to various intervals of time, including a single lesson, a unit, an entire year, or multiple years. Different variables in the curriculum framework might be easier to examine when particular time spans are considered. In the description of the framework, there are points at which the term *curriculum* most naturally refers to a span of an entire year of instruction; at other points, the discussion more easily focuses on a single lesson. Usiskin (2003) uses the term *curriculum size* to refer to different chunks of curriculum, ranging from the single problem to the unit, to the year, to a school's complete mathematics curriculum. When examining the enacted curriculum, researchers must make choices regarding the impact that focusing on different spans of time or size chunks will have on what they study and the claims that will be warranted in their work.

Consideration of organizational levels is more straightforward in its effect on the entire phenomenon under study, but presents additional complexities. Our framework of the enacted curriculum places a spotlight on the classroom level, focusing on the activities of a single teacher in relation to a range of external influences and resources. In school systems, however, teachers are nested in schools. Typically in public school systems, schools are nested in districts, which may be further nested in states or provinces or nations. It is quite possible that many of the variables would look slightly different when considered at varying organizational levels. The designated curriculum, for example, is generally most pronounced and influential at the district level (for schools situated in districts). At the same time, it is likely that schools will add elements to the designated curriculum by indicating a required text or set of resources that teachers are expected to use, or expectations for pedagogical practices teachers should use in their classroom instruction. For simplicity, our discussion of enacted curriculum assumes that the individual teacher and classroom constitute the unit of analysis, yet it is possible to imagine an intended or enacted curriculum at the school level. A curriculum leader at an individual school might play a role in interpreting and translating the designated curriculum for the entire school, creating a school-level intended curriculum. An understanding of the multiple, possible organizational levels of curriculum enactment allows researchers to examine subtle and nuanced elements of the enactment process as they occur in real, nested settings.

GENERATIVE POSSIBILITIES OF THE FRAMEWORK

We anticipate two primary uses of the conceptual framework for research on the enacted curriculum. First, the conceptual framework can serve an analytic purpose, aiding researchers in taking stock of the knowledge base on the enacted mathematics curriculum. In our discussions, to illustrate its analytic usefulness, we situated within the framework a number of studies conducted over the past 20 years on the enacted mathematics curriculum. These studies span grade ranges, national contexts, and research purposes and methods, but can all be located within the territory laid out in the framework. As researchers investigating the enacted curriculum identify particular constructs or relationships of interest, the conceptual framework can serve as an organizer to combine, contrast, and synthesize the bodies of empirical work.

For generative purposes, the framework has the potential to promulgate additional research to deepen and refine knowledge of the proposed constructs and relationships. Ideally, a comprehensive review of relevant literature would provide the field with a sense of findings that have tentative support needing further validation, or support from a limited set of circumstances requiring study under a greater range of conditions. The framework may also aid the field in identifying assumptions or interpretations in existing research that might be challenged in future studies. Additionally, we hope that using the framework as a basis for reviewing literature will uncover gaps in the knowledge base, pointing to possible relationships that might be studied within the curriculum policy, design, and enactment system.

The framework also provides an opportunity for studies that investigate different constructs and relationships within the enacted mathematics curriculum system to be situated alongside one another to enhance the field's knowledge of curriculum as it evolves throughout the system depicted in the framework. For instance, findings from studies of state policies that may influence the enacted curriculum can be put alongside studies of what transpires in classrooms to offer a fuller picture of the connections that the framework posits between these two parts of the education system.

NOTES

1. Initial work conceptualizing this domain was undertaken by a team of researchers at the Conference on Research on the Enacted Mathematics Curriculum in November 2010. The team included Kathryn Chval, Marta Civil, Cassie Freeman, Dan Heck, Beth Herbel-Eisenmann, Mary Ann Huntley, Karen King, Janine Remillard, Jo Ellen Roseman, Jeff Shih, Deborah Spencer, and Mary Kay Stein.
2. The Third International Mathematics and Science Study (TIMSS) used a systemic model of educational opportunity developed for the Second International Study of Mathematics (McKnight, 1979).

REFERENCES

Ben-Peretz,, M. (1990). *The teacher-curriculum encounter: Freeing teachers from the tyranny of texts*. Albany, NY: State University of New York Press.

Boaler, J., & Greeno, J. (2000). Identity, agency, and knowing in mathematics worlds. In J. Boaler (Ed.), *Multiple perspectives on mathematics teaching and learning* (pp. 171-200). Westport CT: Ablex.

Boaler, J., & Staples, M. (2008). Creating mathematical futures through an equitable teaching approach: The case of Railside School. *Teachers College Record, 110*(3), 608-645.

Cal, G., & Thompson, D. R. (2014). The enacted curriculum as a focus of research. In D. R. Thompson & Z. Usiskin (Eds.), *Enacted mathematics curriculum: A conceptual framework and research needs* (pp. 1-19). Charlotte, NC: Information Age Publishing.

Coburn, C. E. (2006). Framing the problem of reading instruction: Using frame analysis to uncover the microprocesses of policy implementation in schools. *American Educational Research Journal, 43*(3), 343-379.

Cohen, D. K., & Ball, D. L. (1999). *Instruction, capacity, and improvement*. (CPRE Research Report No. RR-43). Philadelphia: University of Pennsylvania, Consortium for Policy Research in Education.

Cohen, D. K., & Hill, H. C. (2001). *Learning policy: When state education reform works*. New Haven, CT: Yale.

Council of Chief State School Officers, Wisconsin Center for Education Research, & Eleven State Collaborative. (2000, May). *Using data on enacted curriculum in mathematics & science: Sample results from a study of classroom practices and subject content*. Washington, DC: Council of Chief State School Officers.

Dreeben, R. (1968). *On what is learned in schools*. Boston, MA: Addison-Wesley.

Eisenmann, T., & Even, R. (2009). Similarities and differences in the types of algebraic activities in two classes taught by the same teacher. In J. T. Remillard, B. A. Herbel-Eisenmann, & G. M. Lloyd (Eds.), *Mathematics teachers at work: Connecting curriculum materials and classroom instruction* (pp. 152-170). New York, NY: Routledge.

Franke, M. L., Kazemi, E., & Battey, D. (2007). Understanding teaching and classroom practice in mathematics. In F. K. Lester (Ed.), *Second handbook of research on mathematics teaching and learning* (pp. 225-256). Charlotte, NC: Information Age Publishing.

Goodlad, J. I., Klein, F., & Tye, K. A. (1979). The domains of curriculum and their study. In J. I. Goodlad (Ed.), *Curriculum inquiry: The story of curriculum practice* (pp. 43-76). New York, NY: McGraw-Hill.

Goos, M. (2004). Learning mathematics in a classroom community of inquiry. *Journal for Research in Mathematics Education, 35*(4), 258-291.

Gueudet, G., Pepin, B., & Trouche, L. (2011.). *Text to "lived" resources: Mathematics curriculum materials and teacher development*. New York, NY: Springer.

Gueudet, G., & Trouche, L. (2009). Towards new documentation systems for teachers? *Educational Studies in Mathematics, 71*(3), 199-218.

Herbel-Eisenmann, B., Lubienski, S. T., & Id-Deen, L. (2006). Reconsidering the study of mathematics instructional practices: The importance of curricular context in understanding local and global teacher change. *Journal of Mathematics Teacher Education, 9*(4), 313-345.

Herbel-Eisenmann, B. A., & Otten, S. (2011). Mapping mathematics in classroom discourse. *Journal for Research in Mathematics Education, 42*(5), 451-485.

Hiebert, J., Gallimore, R., Garnier, H., Givvin, K. B., Hollingsworth, H., Jacobs, J., Chui, A. M-Y., Wearne, D., Smith, M., Kersting, N., Manaster, A., Tseng, E., Etterbeek, W., Manaster, C., Gonzales, P., & Stigler, J. W. (2003). *Teaching mathematics in seven countries: Results from the TIMSS 1999 Video Study* (NCES 2003-013). Washington, DC: U.S. Department of Education, National Center for Education Statistics.

Hiebert, J., Thomas, P., Carpenter, T., Fennema, E., Fuson, K., Wearne, D., Murray, H., Olivier, A., & Human, P. (1997). *Making sense: Teaching and learning mathematics with understanding*. Portsmouth, NH: Heinemann.

Hill, H. C., & Charalambos, C. Y. (2012). Teacher knowledge, curriculum materials, and quality of instruction: Lessons learned and open issues. *Journal of Curriculum Studies, 44*(4), 559-576.

Hollebrands, K. F. (2007). The role of a dynamic software program for geometry in the strategies high school mathematics students employ. *Journal for Research in Mathematics Education, 38*(2), 164-192.

Jackson, P. W. (1968). *Life in classrooms*. New York, NY: Holt, Rinehart and Winston.

Lloyd, G. M. (1999). Two teachers' conceptions of a reform-oriented curriculum: Implications for mathematics teacher development. *Journal of Mathematics Teacher Education, 2*(3), 227-252.

McKnight, C. C. (1979). Model for the Second Study of Mathematics. In *Bulletin 4: Second IEA study of mathematics.* Urbana, IL: SIMS Study Center.

McLaughlin, M. W. (1990). The Rand change agent study revisited: Macro perspectives and micro realities. *Educational Researcher, 19*(5), 11-16.

National Council of Teachers of Mathematics. (1989). *Curriculum and evaluation standards for school mathematics.* Reston, VA: Author.

National Research Council. (2001). *Adding it up: Helping children learn mathematics.* In J. Kilpatrick, J. Swafford, & B. Findell (Eds), Mathematics Learning Study Committee, Center for Education, Division of Behavioral and Social Science and Education. Washington, DC: National Academy Press.

Raudenbush, S. W. (2008). Advancing educational policy by advancing research on instruction. *American Educational Research Journal, 45*(1), 206-230.

Remillard, J. T. (1999). Curriculum materials in mathematics education reform: A framework for examining teachers' curriculum development. *Curriculum Inquiry, 100*(4), 315-341.

Remillard, J. T., & Bryans, M. (2004). Teachers' orientations toward mathematics curriculum materials: Implications for teacher learning. *Journal for Research in Mathematics Education, 35*(5), 352-388.

Schmidt, W. H., Jorde, D., Cogan, L., Barrier, E., Ganzalo, I., Moser, U., Shimizu, K., Swada, T., Valverde, G. A., McKnight, C. C., Prawat, R. S., Wiley, D. E., Raizen, S. A., Britton, E. D., & Wolfe, R. G. (1996). *Characterizing pedagogical flow: An investigation of mathematics and science teaching in six countries.* Dordrecht, Netherlands: Kluwer.

Shein, P. P. (2012). Seeing with two eyes: A teacher's use of gestures in questioning and revoicing to engage English language learners in the repair of mathematical errors. *Journal for Research in Mathematics Education, 43*(2), 182-222.

Sleep, L., & Eskelson, S. (2012). MKT and curriculum materials are only part of the story: Insights from a lesson on fractions. *Journal of Curriculum Studies, 44*(4), 537-558.

Snyder, B. R. (1971). *The hidden curriculum.* Cambridge, MA: MIT Press.

Snyder, J., Bolin, F., & Zumwalt, K. (1992) Curriculum implementation. In P. W. Jackson (Ed.), *Handbook of research on curriculum* (pp. 402-435). New York, NY: Macmillan.

Sosniak, L. A., & Stodolsky, S. S. (1993). Teachers and textbooks: Materials use in four fourth-grade classrooms. *Elementary School Journal, 93,* 249-275.

Stein, M. K., Grover, B. W., & Henningsen, M. (1996). Building student capacity for mathematical thinking and reasoning: An analysis of mathematical tasks used in reform classrooms. *American Educational Research Journal, 33*(2), 455-488.

Stein, M. K., & Kaufman, J. H. (2010). Selecting and supporting the use of mathematics curricula at scale. *American Educational Research Journal, 47*(30), 663-693.

Stein, M. K., & Kim, G. (2009). The role of mathematics curriculum materials in large-scale urban reform: An analysis of demands and opportunities for teacher learning. In J. T. Remillard, B. A. Herbel-Eisenmann, & G. M. Lloyd (Eds.), *Mathematics teachers at work: Connecting curriculum materials and classroom instruction* (pp. 37-55). New York, NY: Routledge.

Stein, M. K., & Lane, S. (1996). Instructional tasks and the development of student capacity to think and reason: An analysis of the relationship between teaching and learning in a reform mathematics project. *Educational Research and Evaluation, 2,* 50-80.

Stein, M. K., Remillard, J. T., & Smith, M. S. (2007). How curriculum influences student learning. In F. K. Lester (Ed.), *Second handbook of research on mathematics teaching and learning* (pp. 319-369). Greenwich, CT: Information Age Publishing.

Stigler, J. W., Gonzales, P., Kawanaka, T., Knoll, S., & Serrano, A. (1999, February). *The TIMSS Videotape classroom study: Methods and findings from an exploratory research project on eighth-grade mathematics instruction in Germany, Japan, and the United States.* (NCES 99-074). Washington, DC: U.S. Department of Education.

Tarr, J. E., Reys, R. E., Reys, B. J., Chávez, Ó., Shih, J., & Osterlind, S. J. (2008) The impact of middle-grades mathematics curricula and the classroom learning environment on student achievement. *Journal for Research in Mathematics Education, 39*(3), 247-280.

Usiskin, Z. (2003). The integration of the school mathematics curriculum in the United States: History and meaning. In S. A. McGraw (Ed.), *Integrated mathematics: Choices and challenges* (pp. 19-20). Reston, VA: National Council of Teachers of Mathematics.

Valverde, G. A., Bianchi, L. J., Wolfe, R. G., Schmidt, W. H., & Houang, R. T. (2002). *According to the book: Using TIMSS to investigate the translation of policy into practice through the world of textbooks.* Dordrecht, Netherlands: Kluwer.

Verschaffel, L., De Corte, E., & Lasure, S. (1994). Realistic considerations in mathematical modeling of school arithmetic word problems. *Learning and Instruction, 4*(4), 273-294.

Walshaw, M., & Anthony, G. (2008). The role of pedagogy in classroom discourse: A review of recent research into mathematics. *Review of Educational Research, 78,* 516-551.

Weiss, I. R., Pasley, J. D., Smith, P. S., Banilower, E. R., & Heck, D. J. (2003). *Looking inside the classroom: A study of K-12 mathematics and science education in the United States.* Chapel Hill, NC: Horizon Research.

Wood, T. L, Nelson, B. S., & Warfield, J. (2001). *Beyond classical pedagogy: Teaching elementary school mathematics.* Mahwah, NJ: Erlbaum.

Yackel, E., & Cobb, P. (1996). Sociomathematical norms, argumentation, and autonomy in mathematics. *Journal for Research in Mathematics Education, 27,* 458-477.

CHAPTER 7

RECOMMENDATIONS FOR GENERATING AND IMPLEMENTING A RESEARCH AGENDA FOR STUDYING THE ENACTED MATHEMATICS CURRICULUM

Kathryn B. Chval, Iris R. Weiss, and Rukiye Didem Taylan

This chapter describes activities that took place during the Conference on Research on the Enacted Mathematics Curriculum to begin a dialog about generating and implementing a research agenda for studying the enacted mathematics curriculum. We suggest that initiating this research agenda should be based on the conceptual framework that emerged during the conference. We argue that accomplishing the ideas related to this agenda will require a sustained and coordinated effort with sufficient resources and infrastructure. The rewards from this investment will ultimately improve the development of mathematics curriculum materials and policies that support their use; the teaching and learning of mathematics; and the mathematics education research enterprise as a whole.

INTRODUCTION

The Conference on Research on the Enacted Mathematics Curriculum built on work conducted by the National Science Foundation-supported Center for the Study of Mathematics Curriculum (CSMC), the National Council of Teachers of Mathematics' (NCTM) Linking Research and Practice Research Agenda Project (Arbaugh, Herbel-Eisenmann, Ramirez, Knuth, Kranendonk, & Quander, 2010), and initial work by Remillard (2009) to develop a conceptual model of teacher-curriculum interactions. This conference provided an opportunity to bring together experts in mathematics curriculum, including senior researchers and those who were just embarking on careers in this area, people who had been developing mathematics curriculum materials, and practitioners. Although the organizers of the conference identified multiple purposes, the ultimate goal of the conference was to facilitate the systematic accumulation of knowledge about mathematics curriculum enactment that could guide policy and practice.

The purpose of this chapter is to describe conference activities that focused on research priorities, identify next steps in generating a research agenda, discuss how we build capacity to pursue this research agenda, and argue why this work is significant. We refer to the conceptual framework that resulted from the conference because we consider it a tool that will enable researchers to use common language to situate their research studies and describe the hypothesized relationships they investigate.

BACKGROUND

Not only do teachers interpret curriculum materials differently, they also alter their district-adopted curriculum materials by changing the order of, supplementing, or omitting portions of the materials (Chval, Chávez, Reys, & Tarr, 2009). Additionally, they make some modifications of the curriculum materials in ways that seem consistent with their developers' intent and others that do not seem consistent (Huntley & Chval, 2010). Accordingly, studies of the enactment of mathematics curriculum materials need to consider not only what particular materials are used but also what actually occurs in classrooms that use those materials. As noted by the National Research Council (2004) committee charged with evaluating the effectiveness of various mathematics curricula:

> A standard for evaluation of any social program requires that an impact assessment is warranted only if two conditions are met: (1) the curricular program is clearly specified, and (2) the intervention is well implemented.

Absent this assurance, one must have a means of ensuring or measuring treatment integrity in order to make causal inferences. (p. 100)

The conceptual framework on the enacted mathematics curriculum highlights potentially important components of implementation, components that might be investigated in studies of the integrity of curriculum enactment.

Considerable resources have been devoted to the development of "research-based" mathematics curriculum materials. But the development and adoption of high-quality, research-based materials are not sufficient for improving student learning. Improving student learning requires understanding what it "looks like" to enact a particular set of curriculum materials well (i.e., to promote student learning), and how that varies based on student characteristics. Improving student learning also requires understanding what knowledge and skills teachers need in order to provide that kind of high quality enactment, and the reasons teachers make the decisions they do about which parts of the curriculum materials to implement, in what depth, and in what order.

Remillard and Heck in Chapter 6 of this volume conceptualize official curriculum as composed of three components: curricular goals, content of consequential assessments, and the designated curriculum (*"the set of instructional plans specified by an authorized entity* [for a particular locale], … *be it a district, school or consortium of schools"*). This conceptual model captures the complexity and diversity of studying the factors involved in the enactment of mathematics curriculum materials. Among the teacher attributes that have been hypothesized (and in some cases shown) to be important in mathematics curriculum enactment are depth of teacher content knowledge, extent of teacher familiarity with the curriculum materials, and teacher understanding of the rationale and content of those materials (Remillard, 2005). The context of instruction may also be instrumental in a teacher's curriculum enactment. Potentially important contextual factors at the school/community level include time available for mathematics instruction, school/district/state accountability policies, parental expectations, and the extent of principal support for the instructional approach inherent in a particular set of curriculum materials. Student characteristics are likely to be important as well. For example, Eisenmann and Even (2009) found that a single teacher using the same curriculum materials enacted them quite differently in different classes, suggesting that discipline problems in one class led the teacher to omit some of the meta-level activities that were viewed as integral to promoting understanding of a concept (see also the description by Huntley and Heck in Chapter 2 of this volume).

Individual studies have targeted different components and relationships captured in the conceptual model. For example, some research efforts (Harwell, Post, Maeda, Davis, Cutler, & Kahan, 2007; Huntley, Rasmussen, Villarubi, Sangtong, & Fey, 2000; Post, Harwell, Davis, Maeda, Cutler, & Andersen, 2008) have investigated the relationship between the adoption of curriculum materials and student achievement, without including a detailed observation of teachers' enactment of the materials. In consideration of enactment of curriculum materials, both Harwell et al. (2007) and Post et al. (2008) relied on district personnel and curriculum directors' evaluation of teachers in addition to using classroom observation protocols in a sample of classes in their studies. Both of these studies rated participant teachers' practices as satisfactory with regards to fidelity of implementation without providing further details on teacher practices. These studies aimed to show a relationship between the official curriculum and student achievement without taking into account the transformations that a curriculum undergoes (Stein, Remillard, & Smith, 2007).

Other researchers (Cai, Wang, Moyer, Wang, & Nie, 2011; Tarr, Reys, Reys, Chávez, Shih, & Osterlind, 2008) have considered both the intended curriculum and the enactment of curriculum materials through extensive longitudinal classroom observations when they investigated student learning or achievement in relation to different curriculum materials. For instance, Cai et al. (2011) investigated the extent to which teachers in the study focused on conceptual and procedural understanding in the classroom. During classroom observations, Tarr et al. (2008) documented classroom learning environments focusing on important aspects, such as building instruction on student ideas, encouraging multiple explanations/perspectives, and fostering conceptual understanding. These studies aimed to explore relationships between the official curriculum, enacted curriculum, and student achievement.

Other researchers have investigated factors that influence teachers' enactment of curriculum materials. The focus of these studies was not on student achievement but the relationship between the teacher and the curriculum materials. Superfine (2009) investigated the role of experience on the enactment of reform-based curriculum materials. She found that the experienced teacher preferred to use activities based on her own experience, while the beginning teacher chose to follow curriculum materials closely. Similarly, Christou, Menon, and Philippou (2009) and Isler and Cakiroglu (2009) found that experience played a significant role on teachers' perceptions of the way they use curriculum materials. Remillard and Bryans (2004) examined teacher orientations toward using curriculum materials and their beliefs about teaching and learning while they enacted curriculum materials. Drake and Sherin (2006) interviewed teachers about how they would use and adapt curriculum materials prior to instruction, observed lessons, and then interviewed teachers after they taught these lessons. By exploring the relationship between the intended and the enacted curriculum, Drake and Sherin found different patterns of adaptation of the curriculum materials for different teachers. Jamieson-Proctor and Bryne (2008) explored the impact of both contextual or external factors and beliefs on teachers' enactment. One-third of teachers in this study reported use of curriculum materials because it was expected by parents. Overall, the body of work by these researchers reflects the influence of both classroom factors and external factors on the relationship between the intended curriculum and the enacted curriculum.

Some studies have explored issues of enacting curricula at large scale. For instance, Jorgensen and Perso (2012) discussed issues of equity and teacher quality in the context of enactment of national curriculum in Australia. Similarly, Nyaumwe, Ngoepe, and Phoshoko (2010) identified difficulties of implementation of the national curriculum in South Africa and discussed the role of a student examination system on teachers' enactment patterns. On a smaller scale, Stein and colleagues (Stein & Coburn, 2008; Stein & Kaufman, 2010) investigated influences of district and school policies on teachers' use of curriculum materials. Additionally, Stein and Kaufman (2010) compared implementation of two elementary curricular programs, investigating factors that helped to explain differences in enactment across 48 different teachers. This study not only investigated how teacher characteristics and use of curriculum materials influenced each other, but also considered how district and administrative personnel provided support for teacher learning.

These studies reflect the relationship between an official curriculum and an enacted curriculum as well as the forces that mediate this relationship. However, even with this growing body of work, progress has been limited by a lack of consensus in the field about what aspects of curriculum enactment might be most important in predicting student learning, what teacher and contextual attributes are potentially important for curriculum enactment, and how the various constructs can be defined and measured. As a result, it has been difficult to accumulate knowledge about either the determinants or the impacts of the enacted curriculum. In the following section, we describe the conference activities that facilitated the discussion about moving the field forward in generating research priorities in relation to the enacted mathematics curriculum.

CONFERENCE ACTIVITIES

Throughout the conference, the participants completed a series of tasks (i.e., Research Mapping Task, Research Scenario Task, and Research

Design Task) in small groups to try out the emerging conceptual framework and help identify research priorities. The conference organizers designed these tasks to address the diversity of levels of research experience represented at the conference, ranging from doctoral students to very experienced researchers. In addition, participants were asked to read Lester (2010), Stein, Remillard, & Smith (2007), and O'Donnell (2008) prior to the conference to provide common background knowledge and language as they engaged in conversations.

In this section, we illustrate some of the converging ideas of the small groups, though the groups took different viewpoints in their approach to potential research studies on enacted mathematics curriculum. Some groups emphasized important roles of teachers, students, and curriculum materials as well as impact of different sources regarding the research on enacted curriculum. Other groups focused on ideas related to teachers, including teacher knowledge, understanding of curriculum materials and mathematics, authority, decisions, beliefs, and practice (e.g., questioning, assessing, helping students make connections, giving priority to some elements and not others, pacing and flow of the lesson, and maintaining level of cognitive demand). These ideas are depicted on the conceptual model by the relationship between the instructional materials and the teachers' intended curriculum, as well as enacted curriculum (see Figure 7.1).

Still another group of participants focused their conversations on the relationship between enacted mathematics curriculum and student learning as well as factors that influence student learning (see Figure 7.2). During these conversations, the participants considered the investigation of variables such as student-student interactions, student expectations and roles; students' opportunity to think, do mathematics, and communicate; and evidence of student learning.

Still other participants focused on the importance of curriculum materials and were interested in variables such as nature, intent, and philosophy of teacher and student materials; sequence of tasks; adaptations; and purpose and goals (see Figure 7.3). These ideas reflect the part of the framework where we see the influence of curricular goals on the designated curriculum (as defined by Remillard and Heck in Chapter 6), which in turn shape and are shaped by the instructional materials.

Other ideas on enacted curriculum that surfaced during group discussions included issues of equity, mathematics, classroom norms and culture, and the influence of the district or school. These ideas as well as other factors that influence the nature of instructional materials, such as market forces, professional societies, learning research, curricular goals, and developers' vision, were also represented in the framework.

Furthermore, participants noted their tendencies to take an evaluative lens and criticize either the curriculum materials or the teacher. They

Recommendations for Generating and Implementing a Research Agenda 155

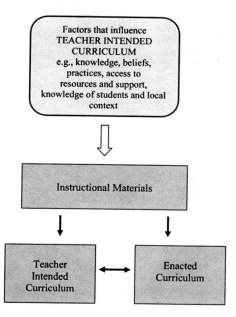

Figure 7.1 Focus on relationships among teachers and enacted curriculum.

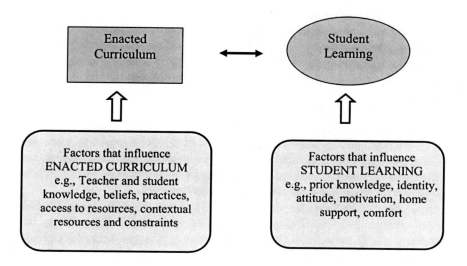

Figure 7.2. Focus on relationships involving students and enacted curriculum.

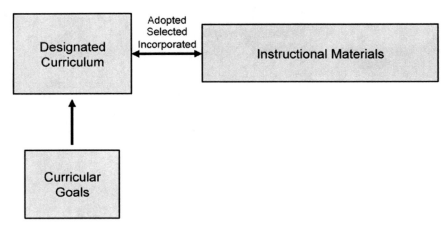

Figure 7.3. Focus on curriculum.

recognized that the teacher's intention may not align with the authors' intentions and they had some concerns about how to distinguish high quality instruction from high fidelity enactment of the curriculum materials. (A poorly-designed set of materials that were enacted with high fidelity would not lead to student learning, so that instruction could not be considered high quality.)

After these initial conversations, the conference facilitators randomly assigned participants to read one of four studies that investigated the enacted mathematics curriculum in order to apply the emerging conceptual framework for research on the enacted mathematics curriculum to concrete examples. These four studies, which represented differences in how researchers studied the enacted curriculum, are summarized in Table 7.1 and described in more detail below.

These four studies were selected by the conference organizers because they represent different aspects of enacted mathematics curriculum (i.e., influences on enacted curriculum, effects of enacted curriculum, and relationships between teachers and enacted curriculum).

Superfine (2009) conducted a case study of two teachers using the *Connected Mathematics* curriculum. This study specifically focused on two teachers' decisions related to the enactment of the mathematics curriculum and their use of teacher guides. Superfine considered four dimensions of the enacted curriculum (i.e., mathematics content, teacher moves, classroom interactions, and use of resources). Through cross-case analysis, she found that these teachers seemed to draw largely from their previous experiences and their own conceptions of mathematics teaching

Recommendations for Generating and Implementing a Research Agenda 157

Table 7.1. Four Studies Related to Curriculum Enactment and Discussed at the Conference

Citation	Brief Description
Superfine, A. C. (2009). The "problem" of experience in mathematics teaching.	This study examined how two sixth grade teachers used the Teacher's Guide (TG) from the *Connected Mathematics Project* curriculum as a resource in making planning and enactment decisions, and factors associated with patterns of TG use.
Manouchehri, A., & Goodman, T. (1998). Mathematics curriculum reform and teachers: Understanding the connections.	This study focused on implementation of four curriculum programs: (a) *Mathematics in Context* (1995), (b) *Sixth through Eighth Mathematics* (1996), (c) *Connected Mathematics Project* (1996), and (d) *Seeing and Thinking Mathematically* (1995). The researchers collected data from 66 middle school mathematics teachers at 12 different school districts over a period of 2 years. The research questions focused on what enhanced and impeded teachers' use of materials, and how they dealt with the challenges they faced.
Brown, S. A., Pitvorec, K., Ditto, C., and Kelso, C. R. (2009). Reconceiving fidelity of implementation: An investigation of elementary whole-number lessons.	This study investigated affordances and limitations of the curriculum in supporting teachers' instructional moves. The researchers analyzed videotapes of 33 first and second grade classroom lessons taught by teachers using the *Math Trailblazers* curriculum in terms of students' opportunities to reason and communicate about mathematics.
Briars, D. J. & Resnick, L. B. (2000). Standards, assessments, and what else? The essential elements of standards-based school improvement.	This study explored effects of implementation of the *Everyday Mathematics* curriculum in U.S. elementary schools. The study involved 57 teachers in 13 schools who were considered weak implementers and 54 teachers in 25 schools as strong implementers. Student achievement was measured via The New Standards Reference Examination and the Iowa Test of Basic Skills.

and learning when making planning and enactment decisions related to mathematical tasks; they did not particularly draw from the teacher guide even though it provided suggestions for the enactment of the curriculum. Additionally, experience seemed to play an important role in the way teachers followed teacher guides. The less-experienced teacher in this study followed the teacher guide more closely than the experienced teacher in the study who preferred to rely more on her experience and own conceptions of teaching mathematics relative to the other teacher.

Manouchehri and Goodman (1998) investigated influences on the nature of enacted curriculum. The results indicated that not only the amount of teaching experience, but also the quality of teaching experience, school context, professional knowledge base and teachers' own personal theories of effective teaching played important roles in the way teachers used curriculum materials in their teaching.

Brown, Pitvorec, Ditto, and Kelso (2009) focused on the relationship between the teachers and the enacted curriculum, specifically paying attention to mathematics content, teacher moves and classroom (student) interactions—three of the dimensions of the enacted curriculum as discussed by Remillard and Heck in Chapter 6. The level of fidelity to the written curriculum materials was associated with following the curriculum materials closely, step by step. The level of fidelity to the authors' intended curriculum was associated with engaging students with the learning opportunities that the textbook authors had in their minds in the design of the textbooks. The authors found that level of fidelity to the written curriculum materials did not predict the level of fidelity to the authors' intended curriculum, and vice versa. Brown et al. argued that the level of fidelity to the written curriculum (literal lesson) differs from the level of fidelity to the authors' intended curriculum during lesson enactments. The results of this study revealed that both the affordances and limitations of the textbooks as well as teacher decisions and choices play a role in the nature of enacted curriculum.

Briars and Resnick (2000) investigated effects of enacted curriculum on student achievement by differentiating different levels of implementation of the same curriculum. In order to determine factors that led to higher student achievement studying the same curriculum, Briars and Resnick compared effects of teacher quality, school context, and student characteristics as well as the nature of implementation. Strong implementers in this study were defined as teachers who used all the components of the curriculum materials and based their instruction on student thinking and explanations. The results revealed that implementation varied across different schools, and that strong implementation was associated with an increase in student achievement.

Research Mapping Task

After the participants read their assigned articles, they met with others assigned to read the same article to complete a research mapping task (shown in Figure 7.4).

A number of ideas surfaced in the sharing across groups that might have implications for a conceptual framework and research agenda. In general, participants indicated that the articles did not provide sufficient description of the written curriculum materials, corresponding teacher materials, or the context in which the study took place. In addition, the participants indicated that the authors typically did not provide a rationale

---

Research Mapping Task

Work with others who were assigned to the same article to answer the following questions about variables/factors reported in the article.

- What variables/factors did the researchers identify and measure about curriculum materials? About classroom practice?
- What other variables/factors did the researchers identify and measure, and how did each of these relate to either curriculum materials, classroom practice, or both?
- Did the study description provide…
    - A good sense of the written curriculum?
    - The rationale for the factors/variables studied in classroom practice?
    - The rationale for the factors/variables in the context that were studied?
- What else would you want to know?

Then discuss the standards of documentation reported in the article:

- What materials were used?
- What are the salient features of the written curriculum?
- What elements of enactment were studied and why?
- What contextual influences were studied and why?

Figure 7.4. Research mapping task.

for the variables that were investigated. The participants would have liked additional information about the students, research instruments and protocols, and research processes (e.g., what was involved in interviews and observations). They posed questions such as:

- Did the teachers participate in professional development?
- What was the nature of the professional development?
- Does the school track students?
- Were teachers observed?
- Why did they select those achievement measures?

During this discussion, participants identified a host of variables to consider in researching the enacted mathematics curriculum, such as teachers' content knowledge, teachers' beliefs about teaching mathematics, teachers' knowledge of written curriculum materials, teachers' experience, the alignment between teacher beliefs and curriculum materials, students' beliefs about learning mathematics, and students' experience (i.e., interaction) with the curriculum materials. The participants acknowledged the need for more studies on teacher learning and knowledge in the context of the enacted mathematics curriculum (e.g., the influence of teachers' knowledge of student learning progressions on enacted curriculum). Additionally, participants expressed a need for studies on how to design professional development programs to support enactment of specific curriculum materials. Similarly, they noted that it is important to consider how teacher educators can help preservice teachers learn how to plan and use curriculum materials. Some participants suggested research related to how content of high-stakes tests influence teachers' intended and enacted curriculum as they believed that standardized testing was a significant force in the current context. Finally, some participants wanted to explore the influence of contextual factors on enacted curriculum, such as time, resources, and norms at both school and district levels.

Research Scenario Task

After participants discussed the Research Mapping Task, the conference organizers used two final tasks to focus the conversation on designing future research. The Research Scenario Task presented in Figure 7.5 involves a specific context; participants were purposively assigned to groups for this task to help ensure diversity in background and experience.

During the discussion of the Research Scenario Task, participants posed questions related to teachers, the assessment, students, curriculum, and context, such as the following:

- Which curriculum materials did the schools use?
- How did the curriculum materials align with the state's standards?
- Does the curriculum (written and enacted) support an opportunity to learn both procedural and conceptual knowledge related to fractions?
- What content related to fractions was taught in previous years?
- Was there a change in curriculum materials?
- What was the nature of instruction in fifth grade, but also at earlier grade levels?

Recommendations for Generating and Implementing a Research Agenda 161

Research Scenario Task

Imagine that 40 states all adopt the same mathematics standards. Now, imagine those same states all administer a common set of assessments to measure students' mathematical understanding. In this purely hypothetical scenario, major emphasis is placed on students' understanding of number and operations with fractions in fifth grade, and the assessment provides a scale score for this topic.

The test results are mixed. Among students whose mathematics performance had been comparable in past years, some students performed very well and others performed poorly. Assume that presently you know nothing more about these students' mathematical learning experiences other than this one assessment score.

In your groups:

1. Brainstorm questions that you would like to ask about the enacted curriculum over the past year to help explain these results.
2. Pick one question. In broad brush strokes, describe what research you might do if you were trying to answer that question for two fifth grade teachers in the same school whose results greatly differed.
3. What would you do differently if you were trying to investigate that question across a large number of classes, e.g., in all 40 states?
4. Based on your list of questions and study descriptions, what additional variables/factors might need to be added to our emerging conceptual framework?

Figure 7.5. Research scenario task.

- What knowledge/skills related to fractions was included on the assessment?
- Were the students actually comparable on important knowledge related to what they ended up studying about fractions in fifth grade?
- What were the characteristics of the teachers (content knowledge as well as teaching and professional development experience), specifically in relation to fractions?

Research Design Task

Prior to the assignment of the Research Design Task (see Figure 7.6), participants identified a preference in terms of grade band (elementary, middle, or secondary) and type of study (small-scale or large scale). Participants met in grade range groups with some groups focusing on small-scale studies, while others worked on large-scale studies, identifying research priorities and sketching out potential designs, including noting instruments to be used and additional instruments needed.

The research questions that participants generated and the variety of methods they suggested helped underscore the idea that a wide variety of possible studies could enrich our understanding of the enacted mathematics curriculum. For example, one group suggested a study of the commonalities and differences among implementation of the same set of curriculum materials across a number of classes, involving videotaping multiple lessons of each class, and relating features of implementation to student learning outcomes. Another group suggested case studies of a small number of teachers, focusing on their interpretation of the curriculum materials they were expected to use, and how that interpretation influenced their implementation of the materials. One group was interested in large-scale survey research to explore the relationship between teacher perceptions of their content understanding and the nature of curriculum enactment. It became apparent that we need a systematic process to consider the variety of options in relation to the conceptual framework and develop a priority research agenda. We outline critical features of that process in the next section.

Research Design Task

- Identify a research question in relation to the enactment of mathematics curriculum.
- Sketch out a potential design, including data sources, instruments, and analysis procedures.
- Consider if your potential study has implications for the emerging conceptual framework and/or the development of additional instruments.

Figure 7.6. Research design task.

NEXT STEPS IN DEVELOPING A RESEARCH AGENDA ON THE ENACTED MATHEMATICS CURRICULUM

The Conference on Research on the Enacted Mathematics Curriculum took place during a time of flux for mathematics curriculum in the United States. Although many countries have a common set of mathematics curriculum standards, the tradition in the United States has been for each state to have its own standards. Reviews of state standards have documented considerable variation among them, and concluded that a student's opportunity to learn important mathematics depends to some extent on where s/he goes to school (Reys, 2006). To help ensure that students throughout the country will experience mathematics in a focused, coherent fashion, the National Governors Association Center for Best Practices and the Council of Chief State School Officers coordinated a multi-year effort to develop common state standards in mathematics. The resulting *Common Core State Standards for Mathematics* (CCSSM) provide a blueprint for the mathematics that students should learn each year as part of their K-12 education to be ready for college and careers (Common Core State Standards Initiative, 2010). In addition to content learning goals, there are eight standards for mathematical practice to be addressed throughout K-12 mathematics education, describing fundamental approaches to, and dispositions towards, learning and doing mathematics. The adoption of the *Common Core State Standards for Mathematics* by more than 40 states, and the funding of two multi-state consortia to develop assessment systems aligned with those standards, has major implications for research on many components of the mathematics education system in the United States, including research on the enacted curriculum.

As the steering committee planned the Conference on Research on the Enacted Mathematics Curriculum, Horizon Research received funding from the National Science Foundation to develop a priority research agenda for investigating the influence of the *Common Core State Standards for Mathematics*. The deliberations at the Enacted Mathematics Curriculum Conference helped inform that process, and in turn, we believe that the process used in developing the CCSSM priority research agenda has implications for next steps in developing a research agenda on the enacted mathematics curriculum. The following sections describe the process for developing a priority research agenda as we envision it, noting which parts have already been accomplished and what else needs to be done.

1. **The research agenda for the enacted mathematics curriculum needs to be considered in the context of a conceptual framework.**

The conceptual framework worked on during the Enacted Mathematics Curriculum Conference and subsequently (see Chapter 6) is a helpful starting point for developing a research agenda on the enacted mathematics curriculum. A subset of participants spent much of their time at the conference working on the development of the conceptual framework, periodically sharing their thinking with, and getting input from, the broader group of conference participants.

In addition to reviewing drafts of the emerging conceptual framework, conference participants tested the utility of the framework for situating a number of existing research studies as well as studies they thought should be included in a research agenda related to the enacted mathematics curriculum. Development of the conceptual framework continued after the conference, with a small group making refinements based on the conference activities and their own thinking. Moreover, it is likely that additional refinements will be made as work on the research agenda continues (e.g., if recommendations for priority studies give rise to variables that need to be added to the framework).

2. **The process for developing a priority research agenda for the enacted mathematics curriculum needs to consider what is already known, and hypothesized, in relation to the conceptual framework.**

Prior to developing a research agenda for studying the enacted mathematics curriculum, it will be important to consider what is known about the various components of the conceptual framework, as well as the relationships among them. Ultimately, the goal is to identify what is known and where there are holes, with particular attention to the components of the conceptual framework that are hypothesized to affect the enacted curriculum most directly.

Findings from existing research can and should be mapped to the conceptual framework for the enacted mathematics curriculum, looking not only at the number of studies that found/did not find evidence of a particular link, but also the strength and generalizability of the evidence, and the important unanswered questions. For example, a review of the literature about deepening mathematics teacher content knowledge (Math and Science Partnership Knowledge Management and Dissemination, 2010) found 14 studies of programs that engaged teachers with challenging mathematics content, with all but one reporting positive impacts on teacher content knowledge. The fact that a number of studies, using a variety of research methods, reached similar conclusions suggests that the finding is a robust one. However, 9 of the 14 studies with positive results were limited to middle grades teachers and 4 of the others also included middle grades teachers along with either elementary or secondary teachers. In a situation such as this, testing the relationship at other grade ranges might well be a priority. In addition, prior research (e.g., Elmore, 2004) has found considerable variation among schools in capacity and resources needed to implement reform. For example, schools that are under a great deal of pressure to improve test scores may have less time and energy to devote to trying to improve learning in important areas not addressed on state assessments. Schools in wealthy communities may have better working conditions, and consequently an easier time than others in attracting and retaining well-prepared mathematics teachers. Consequently, a priority research agenda would likely highlight the need to determine if hypothesized relationships exist in a variety of contexts in order to understand both equity of opportunity and equity of outcomes. Mathematics curriculum researchers can be asked directly about various links in the conceptual model, both to identify studies, including not-yet-published research that was not identified in the search of the literature, and also about their hypotheses regarding the various links in the conceptual model and what kinds of studies would be most important to conduct.

3. **The process for developing a research agenda for the enacted mathematics curriculum needs to include considerable input from the field, involving people with different perspectives on mathematics curriculum and instruction, and on mathematics education research.**

The November 2010 Enacted Mathematics Curriculum Conference provided a foundation for developing a priority research agenda, but much more is needed. The conference provided an opportunity for 78 individuals, including both very experienced and early career mathematics education researchers, to provide input into the initial stages of generating a research agenda. However, neither the conceptual framework nor the summary of what is already known was available; hence, the steering committee recognized that the conference would provide an opportunity to initiate the process to develop a draft of a priority research agenda, but much more work would be needed following the conference.

Others who might not be accustomed to thinking in terms of conceptual models and hypotheses would nevertheless have important contributions to make in developing a priority research agenda. In particular, involving practitioners in setting a research agenda helps ensure that research will be conducted on problems of importance to practice, which in turn should help to bridge the research to practice divide (Burkhardt & Schoenfeld, 2003). For example, district mathematics supervisors could be asked about (1) what mathematics instructional materials are used in the district, and why; (2) the extent to which teachers are using their instructional materials as intended, and how they know; (3) what they think might explain any differences in the enacted mathematics curriculum among teachers, or even among classes taught by the same teachers; and (4) what information they would need to help them make decisions about how to improve mathematics curriculum implementation.

Similarly, curriculum developers could identify differences in the enactment of the curriculum they have seen, or can envision, that would likely affect the effectiveness of the materials; what they have included in teachers' guides to increase the likelihood of implementation as intended; and what else they think needs to be done. Policymakers could be asked about actions they are contemplating in order to improve mathematics teaching and learning, with interviewers probing in areas where research on the enacted curriculum would inform their decisions. Although they may not be stated as research questions involving variables in a conceptual framework, these aspirations, concerns, and insights can and should be translated into that form as part of developing a priority research agenda for the enacted mathematics curriculum.

Summarizing what is known, including the strength of the evidence in support of particular relationships and under what conditions, and identifying practitioner/policymaker information needs, will enable the research agenda process to identify areas where additional research is needed, to test promising findings more fully, and to begin to investigate those new hypotheses that seem most sound on theoretical grounds and most important on practical grounds. As research accumulates, by identifying particular elements of the enacted mathematics curriculum important to student learning, attention can turn to research on how to increase the prevalence of those elements at scale.

4. Convene another conference.

After this initial work is completed and summarized, we suggest another conference that convenes mathematics education researchers, policy researchers, practitioners, representatives of professional organizations, and funders from the federal government and private foundations. To make best use of conference time, participants could be asked to read background documents in advance, for example, the summary of what is already known from research on the enacted mathematics curriculum, the practitioner-identified research needs, and the results of the data collection efforts suggested previously.

To help avoid the possibility of the process generating an overwhelmingly large research agenda, we suggest that some principles be established for setting priorities. For example, addressing practitioners' perceived needs could be given particular weight, with the expectation that *demand-driven* research that answers questions of interest to practitioners would be more likely to be used than *supply-driven* research that addresses questions of interest to researchers. Alternatively, or in addition, identifying improvements that are particularly helpful in narrowing historic achievement gaps could be given priority, given the importance of enabling all students to learn challenging mathematics. Still another alternative would be to prioritize research on elements of the enacted curriculum where improvements could be made at scale with relatively little effort. For example, research could focus on determining which elements of the enacted curriculum are both important for student opportunity to learn and can be most readily improved, (e.g., by providing explanations and examples in teachers' guides) rather than focusing on improvements that can be accomplished only through intensive intervention.

5. **The process for developing a research agenda for the enacted mathematics curriculum needs to consider a variety of types of studies, and needs to include review from people who were not involved in the development of the draft research agenda.**

The steering committee for the Enacted Mathematics Curriculum Conference took the perspective that multiple research approaches are essential. A randomized field trial makes sense to establish causality (e.g., that increasing the use of formative assessment in the enacted curriculum will lead to improved student learning, ruling out the possibility there is something different about the teachers who do and do not use formative assessment that is responsible for the student results). But randomized field trials would not be helpful in understanding why different teachers interpret the same written curriculum material differently.

We anticipate that some version of the process we have outlined here would lead to specification of a multi-faceted research agenda for studying the enacted mathematics curriculum. This research agenda would likely include periodic studies of the status of the enacted mathematics curriculum, focused on the prevalence of elements of the enacted curriculum that evidence suggests are most important in determining opportunity to learn. Relational studies, ranging from small "proofs of concept" studies to broader studies of conditions of effectiveness, would certainly need to be included in a research agenda on the enacted mathematics curriculum. For example, there are studies in the literature that demonstrate teacher knowledge influences or are related to student learning (e.g., Dogan & Adb-El-Khalick, 2008; Hill, Rowan, & Ball, 2005). At the same time, although a small number of studies can demonstrate the viability of a particular relationship, they cannot identify the range of conditions under which a particular relationship exists, and for whom. Broader studies are needed to examine these issues, and to provide possible explanations of why and how a given set of conditions leads to particular outcomes.

As others have noted, alternative sequences in generating knowledge are both possible and helpful (e.g., Heck, Weiss, & Pasley, 2011). Up close and at scale approaches can complement one another in an iterative process, with in-depth case studies, controlled experiments, quasi-experimental designs, national surveys, and so forth, all playing important roles in studying the enacted mathematics curriculum.

For example, in-depth case studies might be helpful in understanding how and why the same designated curriculum and instructional materials wind up with different enactments in different classrooms. What did each teacher intend to do with respect to the mathematics curriculum, and how was that intent influenced by his or her content knowledge, beliefs about student learning, and the available resources? To what extent, and in what ways, was the enacted curriculum different from what the teacher intended, and why? How did professional development on the use of those materials affect the teachers' interpretations and enactments? Subsequent, larger-scale research could investigate whether the nature and extent of the relationship varies depending on teacher background and/or beliefs (e.g., about the relative importance of procedural fluency and conceptual understanding), and the extent to which embedding elements of professional development into teachers' guides affects interpretation and enactment of those and other instructional materials. An alternative sequence might start with a large scale study, perhaps a survey of teachers' interpretations and enactments of one or more sets of instructional materials, followed by case studies to gain insight into the reasons for differences. The draft priority research agenda will need to be reviewed by a large number of people, including people who are not involved in generating the agenda to help ensure that the final product communicates clearly to the broader field. Regardless of which elements of the enacted curriculum are studied and which research approaches are used, it will be important to, over time, address a variety of mathematics content areas, include research across the entire K-12 spectrum, and disaggregate the results to monitor the distribution of high-quality enactment among key demographic subgroups.

6. **To help ensure that the knowledge that is generated provides as comprehensive a picture as possible, the priority research agenda for the enacted mathematics curriculum needs to be accompanied by a plan for ensuring sufficient capacity and resources for the work.**

Recent reviews of the empirical literature in mathematics education have identified substantial deficiencies in the research and a need for the field to adopt more rigorous standards of evidence (e.g., Heck, 2008; Hill & Shih, 2009; National Research Council, 2004). For example, studies may use instruments of questionable validity and reliability or they may use inappropriate units of analysis, such as testing significance based on the number of students involved, even though the treatment was provided at the teacher level.

Web-based seminars and mini-courses focused on research design and implementation offered in conjunction with research conferences (e.g., the NCTM Research Conference[1]) would be particularly cost-effective in developing capacity for research. At the same time, it will be important to encourage researchers both to pursue these kinds of opportunities and to apply what they learn to their research efforts. In our view, the quickest and surest way to move towards a shared set of expectations for rigor in research on the enacted mathematics curriculum, and on mathematics education more broadly, is to provide incentives to the community to do so.

There are often serious methodological weaknesses in the research. This is not surprising given that a survey of mathematics education doctoral programs found that students do not typically enroll in a sufficient number of research methods courses. It is fortunate that, when program representatives were asked about changes they felt were needed to their doctoral programs, the most frequent response was to increase emphasis on preparing students to conduct research (Reys, Glasgow, Teuscher, & Nevels, 2007). Certainly strengthening the initial preparedness of doctoral recipients would be helpful. However, mathematics education researchers, like all professionals, need on-going opportunities for professional development.

The Conference on Research on the Enacted Mathematics Curriculum was designed with capacity-building in mind, deliberately including a mix of doctoral students, recent graduates, and experienced mathematics education researchers. The conference provided an opportunity for early career researchers to learn not only about the role of conceptual frameworks in a body of research, but also about what is entailed in developing these kinds of frameworks. Early career researchers also had an opportunity to interact with some top researchers in mathematics curriculum. In addition, the conference organizers were deliberate in structuring group tasks so they would be both interesting to, and accessible by, researchers with different backgrounds and experience levels, including time for people to go more in depth on topics of their choosing.

By design, early career researchers had ample opportunities for visibility, but with considerable support to help ensure their success. For example, participants had the option of presenting some facet of their work at poster sessions; in addition to the comfort in discussing something about which they were expert, preparing posters and accompanying handouts in advance gave people as much time as they needed to consider what they wanted to say. Similarly, facilitators of the various tasks were asked to make sure everyone had a chance to participate in the conversation, "inviting" the less experienced members of a group to share their thoughts in the small group, and when appropriate with the larger group as well. (In this case, the focus was on research on the enacted mathematics curriculum. However, we believe that whether about some other facet of mathematics education research, or about mathematics education more broadly, conference organizers need to be similarly intentional about building capacity in the field.)

Implementing a research agenda on the enacted mathematics curriculum will require preparing and supporting additional researchers to do this specific work. The Enacted Mathematics Curriculum Conference was a start, but much more will be needed. In particular, we believe that experienced researchers who have pursued research agendas related to the enacted mathematics curriculum will need to work alongside, mentor, and support the next generation of researchers. Doctoral programs will need to prepare future researchers to pursue this line of inquiry.

However, knowledge accumulation depends not only on the quality of individual studies, but also on the extent to which the knowledge that is generated can be aggregated across studies. Moreover, we need more attention to instrumentation (Heck, Chval, Weiss, & Ziebarth, 2012; Hill & Shih, 2009; National Research Council, 2004), and for more complete and consistent documentation of research interventions, contexts, and participants (American Educational Research Association, 2006; Heck & Minner, 2009).

There is much to be gained by the development and use of sound measurement tools. When numerous studies use the same high quality instruments appropriately and well, then we can consider the role of variations in interventions, participants, and contexts to understand different findings, without worrying that the differences are an artifact of instrumentation differences. When research on a particular relationship gets similar results using different high-quality instruments, we can have greater confidence in the findings, without worrying that they are a reflection of a particular way of measuring a construct.

In addition, implementing a priority research agenda on the enacted mathematics curriculum will require stronger relationships with school partners. School personnel are faced with an enormous amount of data to collect, analyze, respond to, and report. Although school personnel are seeking solutions and research-based strategies, they are more reluctant than ever to participate in research that might diminish resources, energy, or flexibility (Chval, Reys, Reys, Tarr, & Chávez, 2006). There is an understandable reluctance to organize students or relinquish precious instructional time for the data gathering often necessary in research studies. There is also an unwillingness to serve as an experimental setting in the present high-stakes, public-accountability environment. Moreover, schools are not always motivated by potential long-term benefits and may be reluctant to wait for results from a research study. Therefore, for educational research to have the opportunity to benefit both researchers and practitioners, it is imperative that researchers identify and implement strategies to address the gap between themselves and practitioners, build trust with school district personnel, provide attractive incentives for research participation, support schools in collecting relevant data, and better inform practitioners about the work they do (Chval et al., 2006).

Implementing a comprehensive research agenda on the enacted mathematics curriculum will require considerably more resources than have been available in recent years. If funding agencies are not able to direct substantial research resources to a particular domain, unless a private foundation decides to do so, progress on a priority research agenda on the enacted curriculum is likely to be quite slow.

Finally, additional communication and coordination structures are needed to support researchers who study the enacted mathematics curriculum. For example, we need a system and infrastructure that will provide an opportunity to accumulate professional knowledge that is storable, shareable (Hiebert, Gallimore, & Stigler, 2002), and tied to the conceptual framework. Rather than individual researchers working in isolation to pursue their individual research agendas, we need a coordinated effort to ensure each study is a contribution to the larger whole.

SIGNIFICANCE

Although there has been increased interest in mathematics curriculum, there has not been a systematic accumulation of knowledge related to mathematics curriculum enactment that can guide policy and practice (Stein, Remillard, & Smith, 2007). Remillard (2009) has identified the lack of explicit theory about the enacted curriculum as a major stumbling block in making more rapid progress in our understanding. She notes:

> The field of research of teachers' use of mathematics curriculum materials lacks a theoretical and conceptual base. As a field, we do not have—or have not been explicit about—theories that underlie and explain the relationships that are central objects of study. As a result, the field has not produced a body of knowledge about the teacher-curriculum material relationship that is generalizable across teachers, materials, or contexts, or that can

inform the work of policymakers, curriculum decision makers, and curriculum material designers in substantive ways. (p. 85)

In addition to a conceptual model along the lines of the one described by Remillard and Heck in Chapter 6 of this volume, substantial progress on a research agenda will require valid and reliable tools, a network of mathematics curriculum researchers with the capacity to conduct and synthesize the proposed research, and significant funding. Accomplishing the ideas represented in this chapter will require a sustained and coordinated effort with sufficient resources and infrastructure. Yet, the rewards from this investment will ultimately improve the development of mathematics curriculum materials and policies that support their use; the teaching and learning of mathematics; and the mathematics education research enterprise as a whole.

NOTE

1. The NCTM Research Pre-Session is now called simply the NCTM Research Conference. As indicated in the introduction to this chapter, NCTM has also periodically held Research Conferences focused on connecting research and practice.

REFERENCES

American Educational Research Association. (2006). Standards for reporting on empirical social science research in AERA publications. *Educational Researcher*, 35(6), 33-40.

Arbaugh, F., Herbel-Eisenmann, B., Ramirez, N., Knuth, E., Kranendonk, H., & Quander, J. (2010). *Linking research and practice: The NCTM research agenda conference report*. Retrieved from http://www.nctm.org/uploadedFiles/Research,_Issues,_and_News/Research/Linking_Research_20100414.pdf#search=%22Linking Research Practice Report Arbaugh%22.

Briars, D. J., & Resnick, L. B. (2000). *Standards, assessments, and what else? The essential elements of standards-based school improvement* (CSE Tech. Rep. No. 528). Los Angeles: University of California, National Center for Research on Evaluation, Standards, and Student Testing.

Brown, S. A., Pitvorec, K., Ditto, C., & Kelso, C. R. (2009). Reconceiving fidelity of implementation: An investigation of elementary whole-number lessons. *Journal for Research in Mathematics Education*, 40(4), 363-395.

Burkhardt, H., & Schoenfeld, A. H. (2003). Improving educational research: Toward a more useful, more influential, and better funded enterprise. *Educational Researcher*, 32(9), 3-14.

Cai, J., Wang, N., Moyer, J. C., Wang, C., & Nie, B. (2011). Longitudinal investigation of the curricular effect: An analysis of student learning outcomes from the LieCal Project in the United States. *International Journal of Educational Research, 50*(2), 117-136.

Christou, C., Menon, M. E., & Philippou, G. (2009). Beginning teachers' concerns regarding the adoption of new mathematics curriculum materials. In J. T. Remillard, B. A. Herbel-Eisenmann, & G. M. Lloyd (Eds.), *Mathematics teachers at work: Connecting curriculum materials and classroom instruction* (pp. 223-244). New York, NY: Routledge.

Chval, K., Chávez, Ó., Reys, B., & Tarr, J. (2009). Considerations and limitations related to conceptualizing and measuring textbook integrity. In J. T. Remillard, B. A. Herbel-Eisenmann, & G. M. Lloyd (Eds.), *Mathematics teachers at work: Connecting curriculum materials and classroom instruction* (pp. 70-84). New York, NY: Routledge.

Chval, K., Reys, R., Reys, B., Tarr, J., & Chávez, Ó. (2006). Pressures to improve student performance: A context that both urges and impedes school-based research. *Journal for Research in Mathematics Education, 37*(3), 158-166.

Common Core State Standards Initiative. (2010). *The standards*. Retrieved from http://www.corestandards.org/the-standards.

Dogan, N., & Abd-El-Khalick, F. (2008). Turkish grade 10 students' and science teachers' conceptions of nature of science: A national study. *Journal of Research in Science Teaching, 45*(10), 1083-1112.

Drake, C., & Sherin, M. G. (2006). Practicing change: Curriculum adaptation and teacher narrative in the context of mathematics education reform. *Curriculum Inquiry, 36*(2), 153-187.

Eisenmann, T., & Even, R. (2009). Similarities and differences in the types of algebraic activities in two classes taught by the same teacher. In J. T. Remillard, B. A. Herbel-Eisenmann, & G. M. Lloyd (Eds.), *Mathematics teachers at work: Connecting curriculum materials and classroom instruction* (pp. 152-170). New York, NY: Routledge.

Elmore, R. (2004). Conclusion: The problem of stakes in performance-based accountability systems. In S. H. Furhman & R. F. Elmore (Eds.), *Redesigning accountability systems for education* (pp. 274-296). New York, NY: Teachers College Press.

Harwell, M. R., Post, T. R., Maeda, Y., Davis, J. D., Cutler, A. L., & Kahan, J. A. (2007). Standards-based mathematics curricula and secondary students' performance on standardized achievement tests. *Journal for Research in Mathematics Education, 38*(1), 71-101.

Heck, D. J. (2008, March). *Applying standards of evidence to empirical research findings: Examples from research on deepening teachers' content knowledge and teachers' intellectual leadership in mathematics and science*. Paper presented at the Annual Meeting of the American Educational Research Association, New York, NY.

Heck, D. J., Chval, K. B., Weiss, I. R., & Ziebarth, S. W. (Eds.). (2012). *Approaches to studying the enacted mathematics curriculum*. Charlotte, NC: Information Age Publishing.

Heck, D. J., & Minner, D. D. (2009). *Codebook for standards of evidence for empirical research*. Chapel Hill, NC: Horizon Research. Retrieved from http://www.mspkmd.net/pdfs/soe.pdf.

Heck, D. J., Weiss, I. R., & Pasley, J. D. (2011). *A priority research agenda for understanding the influence of the common core state standards for mathematics: Technical report*. Chapel Hill, NC: Horizon Research. Retrieved from http://www.horizonresearch.com/reports/2011/CCSSMresearchagenda.technical_report.php

Hiebert, J., Gallimore, R., & Stigler, J. W. (2002). A knowledge base for the teaching profession: What would it look like, and how can we get one? *Educational Researcher, 31*(5), 3-15.

Hill, H. C., Rowan, B., & Ball, D. L. (2005). Effects of teachers' mathematical knowledge for teaching on student achievement. *American Educational Research Journal, 42*(2), 371-406.

Hill, H. C., & Shih, J. (2009). Examining the quality of statistical mathematics education research. *Journal for Research in Mathematics Education, 40*(3), 241-250.

Huntley, M. A., & Chval, K. (2010). Teachers' perspectives on fidelity of implementation to textbooks. In B. Reys, R. Reys, & R. Rubenstein (Eds.), *Mathematics curriculum: Issues, trends, and future directions* (pp. 289-304). Reston, VA: National Council of Teachers of Mathematics.

Huntley, M. A., & Heck, D. J. (2014). Examining variations in enactment of a grade 7 mathematics lesson by a single teacher: Implications for future research on mathematics curriculum enactment. In D. R. Thompson & Z. Usiskin (Eds.), *Enacted mathematics curriculum: A conceptual framework and research needs* (pp. 21-45). Charlotte, NC: Information Age Publishing.

Huntley, M. A., Rasmussen, C. L., Villarubi, R. S., Sangtong, J., & Fey, J. T. (2000). Effects of *Standards*-based mathematics education: A study of the Core-Plus Mathematics Project algebra and functions strand. *Journal for Research in Mathematics Education, 31*(3), 328-361.

Isler, I., & Cakiroglu, E. (2009). Teachers' efficacy beliefs and perceptions regarding the implementation of new primary mathematics curriculum. In V. Durand-Guerrier, S. Soury-Lavergne, & F. Arzarello (Eds.), *Proceedings of the Sixth Congress of the European Society for Research in Mathematics Education* (pp. 1704-1713). Lyon, France.

Jamieson-Proctor, R., & Bryne, C. (2008). Primary teachers' beliefs about the use of mathematics textbooks. In M. Goos, R. Brown, & K. Makar (Eds.), *Proceedings of the 31st Annual Conference of the Mathematics Education Research Group of Australasia* (pp. 295-302). Brisbane, Australia.

Jorgensen, R., & Perso, T. (2012). Equity and the Australian curriculum: Mathematics. In B. Atweh, M. Goos, R. Jorgensen, & D. Siemon (Eds.), *Engaging the Australian national curriculum: Mathematics. Perspectives from the field* (pp. 115-133). Retrieved from http://www.merga.net.au/sites/default/files/editor/books/1/Chapter%206%20Jorgensen.pdf.

Lester, F. K., Jr. (2010). On the theoretical, conceptual, and philosophical foundations for research in mathematics education. In B. Sriraman & L. English

(Eds.), *Theories of mathematics education, advances in mathematics education* (pp. 67-85). Berlin-Heidelberg: Springer-Verlag.

Manouchehri, A., & Goodman, T. (1998). Mathematics curriculum reform and teachers: Understanding the connections. *Journal of Educational Research, 92*(1), 27-41.

Math and Science Partnership Knowledge Management and Dissemination. (2010). *What we know about deepening teachers' content knowledge: Engaging teachers with challenging mathematics and science to deepen their content knowledge, Research on engaging teachers with challenging mathematics content.* Retrieved from http://www.mspkmd.net/index.php?page=01_2c

National Research Council. (2004). *On evaluating curricular effectiveness: Judging the quality of K-12 mathematics evaluations.* Washington, DC: The National Academies Press.

Nyaumwe, L. J., Ngoepe, M. G., & Phoshoko, M. M. (2010). Some pedagogical tensions in the implementation of the mathematics curriculum: Implications for teacher education in South Africa. *Analytical Reports in International Education, 3*(1), 63-75.

O'Donnell, C. L. (2008). Defining, conceptualizing, and measuring fidelity of implementation and its relationship to outcomes in K-12 curriculum intervention research. *Review of Educational Research, 78*(1), 33-84.

Post, T. R., Harwell, M. R., Davis, J. D., Maeda, Y., Cutler, A., & Andersen, E. (2008). *Standards*-based mathematics curricula and middle-grades students' performance on standardized achievement tests. *Journal for Research in Mathematics Education, 39*(2), 184-212.

Remillard, J. T. (2005). Examining key concepts in research on teachers' use of mathematics curricula. *Review of Educational Research, 75*(2), 211-246.

Remillard, J. T. (2009). Considering what we know about the relationship between teachers and curriculum materials. In J. T. Remillard, B. A. Herbel-Eisenmann, & G. M. Lloyd (Eds.), *Mathematics teachers at work: Connecting curriculum materials and classroom instruction* (pp. 85-92). New York, NY: Routledge.

Remillard, J. T., & Bryans, M. B. (2004). Teachers' orientations toward mathematics curriculum materials: Implications for teacher learning. *Journal for Research in Mathematics Education, 35*(5), 352-388.

Remillard, J. T., & Heck, D. J. (2014). Conceptualizing the enacted curriculum in mathematics education. In D. R. Thompson & Z. Usiskin (Eds.), *Enacted mathematics curriculum: A conceptual framework and research needs* (pp. 121-148). Charlotte, NC: Information Age Publishing.

Reys, B. J. (Ed.). (2006). *The intended mathematics curriculum as represented in state-level curriculum standards: Consensus or confusion?* Charlotte, NC: Information Age Publishing.

Reys, R., Glasgow, R., Teuscher, D., & Nevels, N. (2007, November). Doctoral programs in mathematics education in the United States: 2007 Status Report. *Notices of the AMS*.

Stein, M. K., & Coburn, C. E. (2008). Architectures for learning: A comparative analysis of two urban districts. *The American Journal of Education, 114*(4), 583-626.

Stein, M. K., & Kaufman, J. H. (2010). Selecting and supporting the use of mathematics curricula at scale. *American Educational Research Journal, 47*(3), 663-693.
Stein, M. K., Remillard, J., & Smith, M. S. (2007). How curriculum influences student learning. In F. Lester (Ed.), *Second handbook of research on mathematics teaching and learning* (pp. 319-370). Charlotte, NC: Information Age Publishing.
Superfine, A. C. (2009). The "problem" of experience in mathematics teaching. *School Science and Mathematics, 109*(1), 7-19.
Tarr, J. E., Reys, R., Reys, B., Chávez, Ó., Shih, J., & Osterlind, S. (2008). The impact of middle school mathematics curricula and the classroom learning environment on student achievement. *Journal for Research in Mathematics Education, 39*(3), 247-280.

POSTSCRIPT

Zalman Usiskin

A classroom teacher plays many roles, including the following often cited by teacher educators: a manager of both students and learning; an instructor, imparting either as a "sage on the stage" or "a guide on the side"; a coach, providing incentives and imploring students to learn; an evaluator, giving feedback and providing grades; a reporter, indicating to a child's parents or guardians how their charge is proceeding; a counselor, helping students to navigate the world. In addition to these roles, many teachers also consciously act as a curriculum developer, creating activities, lessons, and/or units for their students. Some teachers are on textbook evaluation committees, in which they play a role that can influence classrooms other than their own. None of these roles should surprise anyone in contact with education.

Less obvious is that a teacher is an interpreter of curriculum, a person who takes an existing curriculum and modifies it either consciously or unconsciously to fit the constraints of the teacher's own knowledge, the students, the time available for instruction, and the resources available in the school. Though not so obvious, this role of the teacher as a person who *enacts* an existing curriculum is widely recognized, and provides one reason that state and local guidelines for the teaching of a subject such as mathematics are often written with great detail. One of the purposes of guidelines is to control the breadth of the interpretations a teacher might have of the curriculum as found in official documents or in textbooks or other materials.

A *curriculum* is a prescription of content we wish learned in terms of the students we want to learn that content and the time we want them to learn it—scope, population, and timing. The enactment of a curriculum is not the same as the process of instruction, though obviously it is through instruction that a curriculum is delivered, and some curricula include specific instructional directions.

In describing the process by which mathematics is interpreted in the classroom, I have found it helpful to think of a teacher as a bus driver navigating the world of mathematics, with students as passengers. There are innumerable paths that can be taken through any part of this world. A written curriculum, either in the form of school/district guidelines or textbooks and other materials, is a road map, indicating to the teacher and students what there is to see at each place along the way. The teacher can slow the bus down and explain the sites that can be seen from the bus, ask students to leave the bus to explore on their own, or can drive the bus off on a side road, and in these and other ways control what students see and do not see. In so doing, the teacher is determining the opportunities for students to learn.

Curriculum for a teacher has its counterpart in the itinerary for a bus driver. The itinerary the bus driver is expected to follow is typically not created by the driver but by the company that owns the bus or the tour group that has scheduled the bus or perhaps by a governmental agency. This might start with a broad set of guidelines—see this museum, then travel to that place, stop here, go there, and so forth. In the language of the discussions in Chapters 1 and 6, we might call this the *intended* itinerary. This itinerary is usually communicated to the driver through some written material that might be more detailed, much like a textbook. The *implemented* itinerary is likely to be similar but not exactly like the intended itinerary due to factors such as weather, traffic, dawdling passengers, and so forth.

In Chapter 2 of this volume, Huntley and Heck discuss the differences in instruction that can occur even with the same teacher teaching two classes of the same course with similar students on the same day. They use a signature of chaos theory as a mathematical model, namely that a small perturbation in one place can ultimately lead to a major difference in an outcome. In the classes cited by Huntley and Heck, the perturbation is the presence or absence of a warm-up problem. Another perturbation might have been the presence of teacher aides in one of the classes but not the other.

Such perturbations are everyday occurrences both in traveling on a bus and in classrooms. Visibility on some days is greater than others. A person on a bus or a student in a class may be looking down just when there is something to be seen by looking up. The quality of the ride may be affected by who is sitting next to you, or by what happened the last time the bus driver took this route or the teacher taught this lesson.

These effects are not confined to individual lessons. The presence or absence of a particular student on a particular day not only affects what that student might learn but also can influence what others in the class learn. How else can we explain that, when a teacher has two classes that seem to be identical on a school schedule, with students randomly assigned to them, it is not common that the classes perform alike. Even in the same lesson, students vary in their recall of what transpired. The differences in student performance, in the *attained* curriculum, have their counterpart in the varying memories of different people who were on the same bus trip.

These differences in memory are also due to the enactment of bus tours. The bus driver who knows more about the surrounding area is likely to give a better tour than one who does not. The driver who does not know the landscape or the history may misrelate the story of a particular site, or will omit details that might be important and entire areas that might be interesting. The review of studies by Hunsader and Thompson in Chapter 3 shows the potential influence of the enacted curriculum on student achievement and is further supported by Son and Senk's discussion of research in Chapter 4. What is critical about the research they examine is that it occurs inside the classroom. It may explain why studies that have looked at the effects of such variables as number of mathematics courses taken in college have found no effect on student achievement. "Mathematical knowledge for teaching" is not the same as "mathematical knowledge" overall. If a teacher does not feel comfortable with the content, it may be skipped. If the teacher does feel comfortable, then the more mathematics related to what is being taught that the teacher knows, the richer the potential lesson. Bus drivers, too, may have a great deal of knowledge about some sites and little knowledge about others and steer the conversation accordingly.

The complexity of the enactment of a curriculum is represented by the variety of instruments in the database described by Ziebarth, Fonger, and Kratky in Chapter 5 of this volume. A researcher may survey teachers, students, or administrators to get a handle on what is being implemented, but nothing substitutes for observations, interviews, or both observations and interviews. Analysis of the purposes of these instruments shows that measuring enactment of a set of curricular materials may involve more individuals than the classroom teacher.

This also applies to bus tours. As we have noted, whether the bus is being taken on a local tour of a few hours or a tour in other countries lasting many days, the route is unlikely to have been determined by the bus driver. In addition to the need for a bus tour to be attractive to potential riders, there may be political reasons for visiting some sites and not others, financial reasons for stopping at certain rest points, and socio-historical reasons for hearing about events of the distant or nearby past. People on the tour itself may influence where the tour goes by voicing desires or by staying longer at certain stops than others. The managers and designers of tours need to take all of these factors into consideration, and the bus driver needs to fit the enacted tour into their plans and yet satisfy those doing the touring. The complexity of the process parallels the complexity of the framework of Remillard and Heck shown in Figure 6.1 of Chapter 6.

Chapters 5-7 also bring to mind the process of mathematical modeling. Mathematical modeling begins with a real problem that we wish to solve, a problem that can be quite complex. The process begins by simplifying the problem so that it can be treated mathematically. In studying the process of curriculum enactment, one has to break down the problem so that some aspect of the enactment can be measured. We know that we are losing some of the complexity but we hope that our results are applicable to the more complex situation.

One of the ways by which social problems are simplified is to embed them in theoretical constructs. These constructs are simplifications because, in order to treat a large class of diverse curricular implementations, they have to be overly general. Yet they help in understanding the problem because they introduce a common language by which to understand and discuss the problem and because their generality can enable individuals to fit their local solutions into a broader context. A research agenda, following the suggestions of Chval, Weiss, and Taylan in Chapter 7, and the framework of Remillard and Heck would certainly help to clarify the roles of the many variables involved in studying the enacted curriculum.

But, as everyone knows, the devil is in the details, and we are always influenced as much by case studies with their detail as by broader studies that look for commonalities. That tension is certain to continue even as we become more sophisticated in our look at how curriculum moves from its inception to its classroom use. It is too simple to say that we are trying to apply theory to improve practice; we cannot lose sight of the richness of the raw data, the places where curricula and students meet, in helping us to understand the process. Like the bus driver, we must constantly realize that our passengers are a varied lot, and giving each of them the best experience possible is our ultimate goal.

CONFERENCE ON RESEARCH ON THE ENACTED MATHEMATICS CURRICULUM

University of South Florida
November 4-6, 2010

AGENDA

Prior to the Conference

- Slide presentation created with pictures of participants and brief statements about their interests in curriculum, including one or two issues they believe need addressing.
- A copy of one of the following four research articles about the enacted curriculum is placed in each conference packet.
    o Briars, D. J., & Resnick, L. B. (2000). *Standards, assessments, and what else? The essential elements of standards-based school improvement* (CSE Technical Report 528). Center for the Study of Evaluation, National Center for Research on Evaluation, Standards, and Student Testing, University of California, Los Angeles.
    o Brown, S. A., Pitvorec, K., Ditto, C., & Kelso, C. R. (2009). Reconceiving fidelity of implementation: An investigation of

elementary whole-number lessons. *Journal for Research in Mathematics Education, 40*(4), 363-395.
- o Manouchehri, A., & Goodman, T. (1998). Mathematics curriculum reform and teachers: Understanding the connections. *Journal of Educational Research, 92*(1), 27-41.
- o Superfine, A. C. (2009). The "problem" of experience in mathematics teaching. *School Science and Mathematics, 109*(1), 7-19.

Thursday, November 4, 2010

2:00-4:00	**Registration/Slide Show**	
	• Slide presentation with pictures of participants and curriculum statements continuously plays.	
4:15-6:00	**Opening Session**	
	• 4:15-4:30	Welcome, purpose of conference, icebreaker
	• 4:30-4:40	Introduction of opening task
	• 4:40-5:10	Participants view videos and write individual thoughts
	• 5:10-5:45	Small group discussion and task
	• 5:45-5:55	Brief report out
6:00-6:45	**Dinner**	
7:00-8:00	**Second Session**	
	• 7:00-7:30	Sense-making of video task
	• 7:30–8:00	Wrap-up of the day
		Homework assignment to read the study included in the registration packet

Friday, November 5, 2010

8:30-10:00	**Third Session**	
	• 8:30-9:30	Introduction of a simple conceptual framework
	• 9:30-10:00	Discussion of research studies and identification of variables
10:00-10:30	**Break**	
10:30-12:00	**Fourth Session**	
	• 10:30-11:00	Presentation of frameworks in the research literature that are related to components of the enacted curriculum and the need for conceptual frameworks
	• 11:00-12:00	Identifying variables when researching the enacted curriculum
12:00-1:00	**Lunch**	
1:00-2:45	**Fifth Session**	
	• 1:00-2:45	Instruments for studying the enactment of curriculum materials

2:45-3:15	**Break**	
3:15-4:45	**Sixth Session**	
	• 3:15- 4:45	The conceptual framework revisited
	• 4:45-5:15	Gallery walk of work done by groups related to the conceptual framework
5:15-5:45	**Seventh Session**	
	• 5:15-5:45	Poster session of participants' research
6:00-7:00	**Dinner**	

Saturday, November 6, 2010

8:00-9:00	Eighth Session	
	• 8:00-9:30	Small group activities to discuss priority research on the enactment of curriculum materials
9:30-9:45	**Break**	
9:45-11:20	**Ninth Session**	
	• 9:45-11:00	Small groups continue work on research agenda
	• 11:00-11:20	Closing thoughts and next steps

LIST OF PARTICIPANTS AND AFFILIATIONS AT TIME OF CONFERENCE

* Indicates members of the Steering Committee

	Name	Institution
	Sarah Bleiler	University of South Florida
	Kelley Buchheister	University of Missouri
	Laura Burr	University of New Mexico
	Gabriel Cal	University of South Florida
	Jeffrey Choppin	University of Rochester
*	Kathryn Chval	University of Missouri
	Marta Civil	University of Arizona & CEMELA
	Jon D. Davis	Western Michigan University
	Zandra de Araujo	University of Georgia
	Richelle C. R. Dietz	North Carolina State University
	Barbara J. Dougherty	Iowa State University
	Alden J. Edson	Western Michigan University
	Sarah Enoch	Portland State University
	Anne Estapa	University of Missouri
	Nicole Fonger	Western Michigan University

(List continues on next page)

PARTICIPANT LIST

	Cassie Freeman	University of Chicago
	Maisie L. Gholson	University of Illinois at Chicago
	Funda Gonulates	Michigan State University
	Brian Greer	Portland State University & CLT-West
	Douglas Grouws	University of Missouri
*	Daniel Heck	Horizon Research, Inc.
	Beth Herbel-Eisenmann	Michigan State University
	Sarah J. Hicks	Rockhurst University
	Christian Hirsch	Western Michigan University
	Krista Holstein	North Carolina State University
*	Mary Ann Huntley	Cornell University
	Lisa Kasmer	Grand Valley State University
	Ok-Kyeong Kim	Western Michigan University
	Karen D. King	New York University
	Darlene E. Kohrman	Kalamazoo Valley Community College & Michigan State University
	Glenda Lappan	Michigan State University
	Lesley F. Leach	University of Texas at Austin, Charles A. Dana Center
	Vena M. Long	University of Tennessee & ACCLAIM
	James Lynn	University of Illinois at Chicago
	Lorraine M. Males	Michigan State University
	Melissa McNaught	University of Iowa
	F. Joseph Merlino	21st Century Partnership for STEM Education
	Helena P. Miranda	Florida Gulf Coast University (evaluator)
	Rebecca Mitchell	Boston College
	Swapna Mukhopadhyay	Portland State University & CLT-West
	Courtney Nelson	Horizon Research, Inc.
	Jill Newton	Purdue University
	Travis A. Olson	University of Nevada, Las Vegas
	Samuel Otten	Michigan State University
	Elizabeth (Betty) Phillips	Michigan State University
	Shagufta Raja	Phillip O'Berry High School CMS & University of North Carolina at Charlotte
*	Janine Remillard	University of Pennsylvania
	Barbara Reys	University of Missouri
	Robert Reys	University of Missouri
	Kimberly Rimbey	Arizona State University

Participant List

	Jo Ellen Roseman	Project 2061/AAAS
	Mollie Rudnick	CEMSE - University of Chicago
	Derrick Saddler	University of South Florida
	Cynthia L. Schneider	University of Texas at Austin, Charles A. Dana Center
	Christina Schwarz	Michigan State University & CCMS
	Ruthmae Sears	University of Missouri
*	Sharon L. Senk	Michigan State University
	Jeffrey Shih	University of Nevada, Las Vegas
	Laura M. Singletary	University of Georgia
	Ji-Won Son	University of Tennessee at Knoxville
	Deborah Spencer	K-12 Mathematics Curriculum Center, EDC
	Michael D. Steele	Michigan State University
	Mary Kay Stein	University of Pittsburgh
	James E. Tarr	University of Missouri
	R. Didem Taylan	University of Missouri
	Cynthia Taylor	University of Missouri
	Megan Westwood Taylor	Harvard University
	Amanda Thomas	University of Missouri
*	Denisse R. Thompson	University of South Florida
	Barbara Trujillo	Office of Education Accountability, New Mexico
	Zalman Usiskin	University of Chicago
	Sarah vanIngen	University of South Florida
	Eugenia Vomvoridi-Ivanovic	University of South Florida
	Tad Watanabe	Kennesaw State University
*	Iris R. Weiss	Horizon Research, Inc.
	Linda Dager Wilson	Project 2061/AAAS
*	Steven W. Ziebarth	Western Michigan University
	Barbara Zorin	University of South Florida

ABOUT THE AUTHORS

Gabriel Cal is Lecturer in Education & Arts at the University of Belize. He received his PhD in Curriculum and Instruction with an emphasis in mathematics education from the University of South Florida, focusing his research on the alignment of national curricular goals, assessments, and curriculum materials at the middle grades in Belize. Dr. Cal has been interested in and involved with teacher training in Belize since 1995.

Kathryn B. Chval is the Associate Dean for Academic Affairs in the College of Education at the University of Missouri in addition to being an Associate Professor of Mathematics Education. Dr. Chval is also the principal investigator for the *Facilitating Latinos' Success in Mathematics* Project and co-principal Investigator for the *Center for the Study of Mathematics Curriculum* and the *Researching Science and Mathematics Teacher Learning in Alternative Certification Models* Project, all funded by the National Science Foundation. Prior to joining the University of Missouri, Dr. Chval was the acting section head for the Teacher Professional Continuum Program in the Division of Elementary, Secondary and Informal Science Division at the National Science Foundation. Dr. Chval's research interests include effective preparation models and support structures for teachers across the professional continuum; effective elementary teaching of underserved populations, especially English language learners; and curriculum standards and policies.

Nicole L. Fonger is a research associate at the Friday Institute for Educational Innovation in the College of Education at North Carolina State University. She earned her BA in mathematics from the University of Saint Thomas in Saint Paul, Minnesota. For her graduate coursework, she earned her MA in mathematics, MA in mathematics education, and PhD in mathematics education from Western Michigan University in Kalamazoo. For her dissertation, she investigated the characterization and support of algebra students' change in representational fluency in a CAS and paper-and-pencil environment. She has experience working with a curriculum development team, and conducting curriculum analyses using a diversity of frameworks and reconciling processes. She has also conducted classroom teaching experiments, engaged teachers in collaborative research, and conducted task-based interviews with students. She has taught at the secondary school and university levels, with a focus on mathematics content courses for elementary and secondary preservice mathematics teachers, and has worked with inservice teachers in face-to-face and online professional development. Her ongoing research pursuits include effective design for linking research and practice on curriculum and policy issues, the design for and implementation of computing technology for teaching and learning mathematics, and supporting learning trajectory based instruction.

Daniel J. Heck is Senior Researcher and Partner at Horizon Research, Inc. in Chapel Hill, North Carolina. Dr. Heck's research interests include mathematics teachers' professional learning, the role of mathematics curriculum materials in classroom instruction, and strategic leadership of education improvement efforts. Dr. Heck served as Principal Investigator of *Developing a Research Agenda for Understanding the Influence of the Common Core State Standards for Mathematics*, and is currently Co-PI of *Fostering Mathematics Success for English Language Learners*, *All Included in Mathematics*, and the Evaluation of the National Science Foundation's *Research and Evaluation in Engineering and Science Education Program*.

Patricia D. Hunsader is Assistant Professor and Director of the Center of Partnerships for Arts-Integrated Teaching (PAInT) at the University of South Florida, Sarasota-Manatee (USFSM). She has taught graduate and undergraduate mathematics methods courses for over ten years, and her research interests include mathematics curriculum, arts integration and classroom assessment. She has published and presented her research both nationally and internationally, most recently at the Twelfth International Congress on Mathematical Education in Seoul, Korea, and co-authored the 7th edition of *Mathematics: A Good Beginning* in 2013. Dr. Hunsader was granted the Outstanding Professor Award from USFSM in 2011 and in 2013, and was named a STaR (Service, Teaching, and Research) Fellow, a project funded by the National Science Foundation, in 2011. In 2012, she received the Outstanding Teaching Award from USFSM.

Mary Ann Huntley holds the position of Senior Lecturer of Mathematics and Director of Mathematics Outreach and K-12 Education Activities at Cornell University. She is a Research Associate with the Center for the Study of Mathematics Curriculum, housed at the University of

Missouri. She has received numerous awards for her scholarly work, including a National Academy of Education/Spencer Postdoctoral Fellowship (2003) and the AACTE Outstanding Dissertation Award (1997). Building on her experience examining and analyzing mathematics classroom practice at all grade levels in schools across the United States, her primary research interest involves investigating the relationships between mathematics curricula, teaching, and students' learning, especially at the middle- and high-school levels.

James L. Kratky is a PhD student in the Department of Mathematics at Western Michigan University. Before beginning his graduate studies, he taught high school mathematics classes, ranging from Algebra 1 to AP Statistics. As a graduate student, he has participated in the data collection and analysis of a National Science Foundation funded study that examined the influence of a preservice mathematics teacher methods sequence on teachers' practice, served as a student research associate under the Center for the Study of Mathematics Curriculum, worked as an assistant under the *Transition to College Mathematics and Statistics Project*, and taught undergraduate mathematics courses. During 2012, he served as an editor on the *Proceedings of the 34th Annual Meeting of the North American Chapter of the International Group for the Psychology of Mathematics Education*. Currently, he is investigating ways in which teachers guide and shape students' use of mathematical technologies in the classroom.

Janine T. Remillard is Associate Professor of Mathematics Education at the University of Pennsylvania's Graduate School of Education. Dr. Remillard's research interests include teachers' interactions with mathematics curriculum materials, mathematics teacher learning in urban classrooms, and locally relevant mathematics instruction. She is one of the primary faculty in Penn-GSE's urban teacher education program and is co-editor of the volume, *Mathematics Teachers at Work: Connecting Curriculum Materials and Classroom Instruction*. She is Principal Investigator of two NSF-funded studies: *Improving Curriculum Use for Better Teaching* and *Learning About New Demands in Schools: Considering Algebra Policy Environments*. Dr. Remillard chairs the U.S. National Commission on Mathematics Instruction, a commission of the National Academy of Sciences.

Sharon L. Senk is Professor in the Program in Mathematics Education and the Department of Mathematics at Michigan State University. She taught mathematics to elementary and secondary teachers in Colombia for 2 years, and to high school students in Massachusetts for 12 years. In 1983 she received a PhD with an emphasis on mathematics curriculum from the University of Chicago, and later was an Assistant Professor at

Syracuse University and a Research Associate at the University of Chicago. Her interests include learning and teaching secondary school mathematics, especially as pertaining to reasoning and proof, curriculum development and research, and the mathematical preparation of teachers. For two decades she served as Co-Director of the Secondary Component of the University of Chicago School Mathematics Project. More recently, she was Co-Director of the 17-country Teacher Education and Development Study, a comparative study of teacher education under the auspices of the International Association for the Evaluation of Educational Achievement. Her current research is investigating the preparation of secondary school teachers for teaching algebra.

Ji-Won Son is an Assistant Professor of Mathematics Education at the University at Buffalo—The State University of New York. Prior to this position, she spent five years as an Assistant Professor at the University of Tennessee at Knoxville. She received her doctorate from Michigan State University in 2008 with an emphasis in mathematics education. Previously, she spent four years teaching elementary and middle school students in South Korea. Her main research interests include: mathematics textbook analysis; inservice teachers' textbook use, especially as it relates to the cognitive demand on student thinking; elementary and secondary preservice teachers' knowledge in relation to their beliefs about teaching; and international comparative studies. She is interested in continuing her investigations of inservice teachers' textbook use through professional development programs; she is particularly interested in finding ways to help teachers increase their content understanding and improve their teaching practice.

Rukiye Didem Taylan completed her PhD in the Mathematics Education program in the department of Learning, Teaching and Curriculum at the University of Missouri. A Fulbright Scholarship recipient, she received her MS in Mathematics Education at Teachers College, Columbia University. Her research interests include teacher education and professional development, teachers' use of curriculum, and teacher noticing. She has taken a position at Çanakkale Onsekiz Mart University in Turkey.

Denisse R. Thompson is Professor of Mathematics Education in the College of Education at the University of South Florida (USF) in Tampa. She has taught at the middle school, high school, and community college levels and joined the faculty at USF in 1990. Her scholarly interests include curriculum development and research, literacy in mathematics, and the integration of culture and literature into the teaching of mathematics. She has been involved with the University of Chicago School Mathematics

Project for over 25 years—as an author, editor, and most recently as Director of Evaluation for the Third Edition Secondary materials. She has authored or co-authored 18 books, over 25 book chapters, and over 50 journal articles.

Zalman Usiskin is Professor Emeritus of Education at the University of Chicago and the overall director of the University of Chicago School Mathematics Project (UCSMP), a position he has held since 1987. He is also a Co-Principal Investigator of the Center for the Study of Mathematics Curriculum. His research has focused on the teaching and learning of arithmetic, algebra, and geometry, with particular attention to applications of mathematics at all levels and the use of transformations and related concepts in geometry, algebra, and statistics. He is interested in matters related to mathematics curriculum, instruction, and testing; international mathematics education; the history of mathematics education; and educational policy.

Iris R. Weiss is founder and President emeritus of Horizon Research, Inc. (HRI), a contract research firm in Chapel Hill, North Carolina, specializing in science and mathematics education research and evaluation. Before establishing HRI in 1987, Dr. Weiss was senior education research scientist at the Research Triangle Institute. At HRI, she directed periodic *National Surveys of Science and Mathematics Education*, the *Inside the Classroom* national observation study, and evaluations of a wide variety of mathematics and science professional development and systemic reform initiatives. Dr. Weiss was principal investigator of the *Core Evaluation of the Local Systemic Change through Teacher Enhancement Initiative*, co-directed the *Knowledge Management and Dissemination for NSF's Math-Science Partnerships*, developed a priority research agenda for the *Common Core State Standards for Mathematics*, and represented HRI on the leadership team of the Center for the Study of Mathematics Curriculum. Her research interests include professional development for mathematics and science teachers, and going to scale with mathematics and science education reform efforts.

Steven W. Ziebarth is Professor of Mathematics Education in the Department of Mathematics at Western Michigan University. His research interests focus on secondary curriculum, student assessment, and evaluation. Over the past two decades, he has been lead evaluator on several curriculum and professional development projects including the National Council of Teachers of Mathematics' *Project to Implement the Standards in Discrete Mathematics*, the Iowa Local Systemic Change Initiative, the National Science Foundation [NSF] funded *Core Plus Mathematics Project* and Second Edition Revision, the Iowa Important Mathematics and

Powerful Pedagogies Project, and currently the *Transition to College Mathematics and Statistics Project* (TCMS). He is also the Director of the NSF-funded *Assessment for Learning* capacity building project at Western Michigan University, the Principal Evaluator of the NSF-funded *Assessing Teachers' Pedagogical Design Capacity and Mathematics Curriculum Use*, and a member of the leadership team for the Center for the Study of Mathematics Curriculum.

CPSIA information can be obtained at www.ICGtesting.com
Printed in the USA
BVOW04s1206121213

338893BV00003B/36/P